EX LIBRIS
YJB-065655 The Rainbow Fish by Marcus Pfister © 1992 by Nord-Süd Verlag AG

SCANNING AND TRANSMISSION ELECTRON MICROSCOPY
AN INTRODUCTION

SCANNING AND TRANSMISSION ELECTRON MICROSCOPY
AN INTRODUCTION

Stanley L. Flegler
John W. Heckman, Jr.
Karen L. Klomparens

Center for Electron Optics
Michigan State University

Oxford University Press
New York Oxford

Oxford University Press

Oxford New York
Athens Auckland Bangkok Bombay
Calcutta Cape Town Dar es Salaam Delhi
Florence Hong Kong Istanbul Karachi
Kuala Lumpur Madras Madrid Melbourne
Mexico City Nairobi Paris Singapore
Taipei Tokyo Toronto

and associated companies in
Berlin Ibadan

Published by Oxford University Press, Inc.,
198 Madison Avenue, New York, New York 10016-4314

Library of Congress Cataloging-in-Publication Data
Flegler, Stanley L.
Scanning and transmission electron microscopy : an introduction /
Stanley L. Flegler, John W. Heckman, Jr., Karen L. Klomparens.
p. cm.
Originally published: New York : W.H. Freeman, c1993.
Includes bibliographical references and index.
ISBN 0-19-510751-9
1. Scanning electron microscopy. 2. Transmission electron
microscopy. I. Heckman, John William. II. Klomparens, Karen L.
III. Title.
QH212.S3F58 1995 95-39017
502′.8′25—dc20 CIP

1 3 5 7 9 8 6 4 2

Printed in the United States of America
on acid-free paper

CONTENTS

PREFACE

This textbook was written to fulfill a perceived need in the field of electron microscopy for an introductory, interdisciplinary book designed to be used as the sole textbook for a lecture course and as a reference textbook for laboratory courses. Although we sought repeatedly to find a single textbook or a combination of two textbooks appropriate for these courses, we found none of the existing textbooks entirely suitable, even though many are excellent for the students they target.

To the writing of this textbook we bring more than forty years of combined experience in teaching a general upper-division and graduate-level university course intended as a survey course for all university departments. The class is a requirement for enrollment in various laboratory classes (SEM, TEM, EDS) taught at the Center for Electron Optics at Michigan State University. During the past twenty years, more than a thousand students have taken the survey class and subsequently have enrolled in the laboratory classes.

Our goal was to write a textbook to meet the following criteria. (1) The text should cover SEM, TEM, STEM, and EDS. (2) The text should be applicable to the theoretical and applied disciplines of both physical and biological sciences. (3) The text should be written at the introductory level, with appropriate references provided for readers interested in more advanced information. (4) The text should be inexpensive. (5) The text should not include any detailed laboratory procedures, which usually are specific to the laboratory or instructor and tend to change frequently. Specific laboratory instructions are more appropriate in a laboratory manual or in individually prepared handouts. Most electron microscopists have spent years developing and carefully revising particular laboratory exercises that reflect their specific equipment and the disciplines of the students they teach. With this in mind, we have endeavored to provide the theoretical and fundamental principles in a straightforward style, relying on individual instructors to use the protocols and procedures specific to their own laboratory.

We sought to produce a text that used illustrations primarily to demonstrate how electron microscopes and related accessories function and how the various changes in operational parameters can affect the results. To this end, we included only essential illustrations; in our view excessive illustration only increases the price of a book and the nonessential descriptions in the legends only distract the reader.

We have intentionally written an interdisciplinary text. In recent years many of the once-clear divisions between disciplines have become less distinct. A major shortcoming of some texts is that they usually restrict the reader to the techniques in use at that time in the given discipline and therefore do not allow for understanding and using other preparation or analysis techniques.

We wish to thank the many colleagues, graduate teaching assistants, and students who have assisted in the writing of this textbook by providing critical reviews. In particular our thanks go to Dr. Martin Crimp of the Department of Materials Science and Mechanics and Dr. Carl Foiles of the Department of Physics and Astronomy at Michigan State University for providing specific review and assistance with the sections on SEM and TEM diffraction techniques. In addition, we wish to thank the following individuals for providing photographs: R. S. Cornell, M. J. Crimp, J. Everard, C. H. Neilson, V. E. Robertson, J. J. Stout, and J. H. Whallon.

Stanley L. Flegler
John W. Heckman, Jr.
Karen L. Klomparens
June 1993

1

Introduction

The 1986 Nobel Prize in Physics was awarded jointly to Ernst Ruska, Gerd Binnig, and Heinrich Rohrer. Ruska, of the Fritz Haber Institute in West Berlin, was recognized for his work on electron optics and the original design of the transmission electron microscope in the 1930s. Binnig and Rohrer, of the IBM Zurich Research Laboratory, were acknowledged for their design of the scanning tunneling microscope a half century later in the 1980s. In presenting the award, the Nobel Prize Committee noted the significance of electron microscopes in all areas of science, calling them "one of the most important inventions of this century."

Defined very simply, electron microscopy is the science and technology of using an electron beam to form magnified images of specimens. The principal advantage of using electrons, rather than light, to form images is that electrons provide as much as a thousandfold increase in resolving power—i.e., the ability to distinguish fine detail. The resolving power of a modern light microscope is, at best, 200 nm, whereas a transmission electron microscope can resolve detail to approximately 0.2 nm, and a scanning electron microscope to approximately 3 nm for specially designed instruments. The desire to resolve finer and finer detail led to the development of the electron microscope in the 1930s and 1940s. With continuing refinement, the electron microscope has enabled scientists in a variety of disciplines to gather data based on high-resolution images that are not available with any other technique.

Electron microscopy is used in a variety of ways in biology, medicine, and the materials sciences. Examples in the biological sciences include diagnosis of human, animal, and plant diseases; development of disease pathology; study of morphological and developmental aspects of organisms, tissues, and cells; identification of pollen, viruses, bacteriophages, and diatoms; visualization of subcellular components and structures such as DNA; subcellular localization of elements, enzymes, and proteins; in situ hybridization of gene products; and confirmation of biological, physiological, and biochemical data. Examples in the materials and physical sciences include identification of minerals in geological specimens; determination of the crystal lattice structure of such materials as catalysts and layered clays; determination of the structure of composite materials, thin films, and ceramics; fracture analysis and alloy composition in metallurgical samples; localization and quantitation of elements in samples; and confirmation of chemical, metallurgical, and geological data.

Transmission and Scanning Electron Microscopes Produce Unique Images

The two basic types of electron microscope are the transmission electron microscope (TEM, Figure 1.1) and the scanning electron microscope (SEM, Figure 1.2), which produce unique, and often complementary, images and information. Table 1.1 compares selected characteristics of these two electron microscopes and the light microscope. Both the SEM and the TEM have greater resolving capabilities, as well as a wider and higher magnification range, than the light microscope does. Electron microscopes require operation under high-vacuum conditions (see Chapter 3), with the exception of an environmental SEM, which has a high vacuum in most of the instrument but also has a sample chamber that can operate at near atmospheric conditions (see Chapter 5). Specimens for the TEM and SEM often require more extensive preparation than that needed for the light microscope. Proper specimen preparation is critical to image quality, and therefore data acquisition, in electron microscopy. The specific requirements for SEM and TEM specimen preparation are described in Chapters 6 and 7.

A simple analogy can be made between a TEM and a compound light microscope. The bright-field images from both microscopes provide data on the structure of the internal components of the specimen. For each microscope, successful imaging depends on a specimen that is

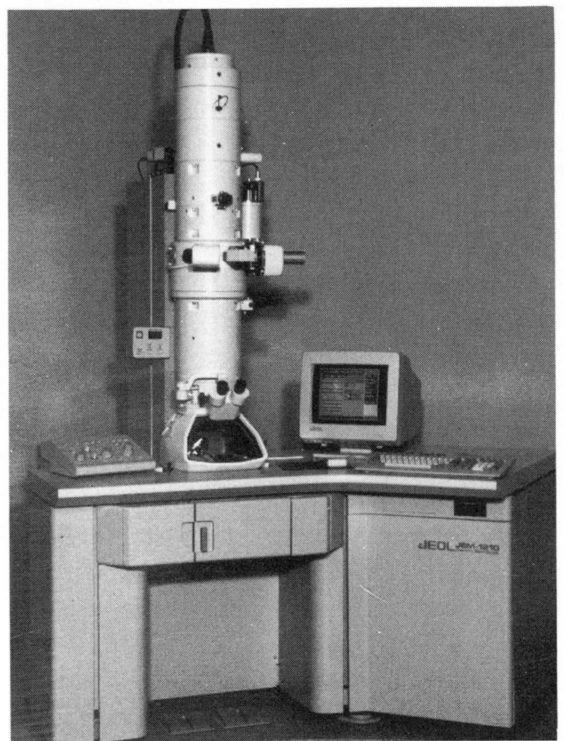

Figure 1.1 Transmission electron microscope. Photo courtesy of Japan Electron Optics Laboratory (JEOL), USA, Inc., Peabody, MA.

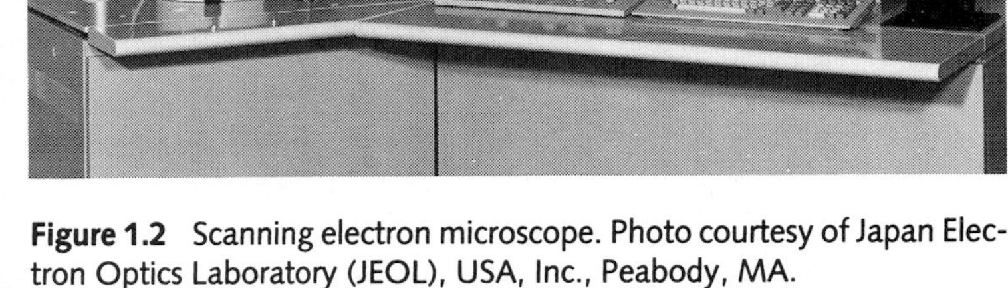

Figure 1.2 Scanning electron microscope. Photo courtesy of Japan Electron Optics Laboratory (JEOL), USA, Inc., Peabody, MA.

thin enough for either the electron beam (TEM) or the light beam (light microscope) to pass through, as well as on the correct operation of the specific imaging and magnifying lenses (see Chapter 2). The TEM can produce a high-resolution image of cellular ultrastructure (Figure 1.3) or of isolated particulates, such as viruses. Although the analogy be-

TABLE 1.1
Comparison of Selected Characteristics of Light and Electron Microscopes

FEATURE	LIGHT MICROSCOPE	TRANSMISSION ELECTRON MICROSCOPE	SCANNING ELECTRON MICROSCOPE
General use	Surface morphology and sections (1–40 μm)	Sections (40–150 nm) or small particles on thin membranes	Surface morphology
Source of illumination	Visible light	High-speed electrons	High-speed electrons
Best resolution	ca. 200 nm	ca. 0.2 nm	ca. 3–6 nm
Magnification range	10–1,000×	500–500,000×	20–150,000×
Depth of field	0.002–0.05 nm (N.A. 1.5)	0.004–0.006 mm (N.A. 10^{-3})	0.003–1 mm
Lens type	Glass	Electromagnetic	Electromagnetic
Image ray-formation spot	On eye by lenses	On phosphorescent plate by lenses	On cathode tube by scanning device

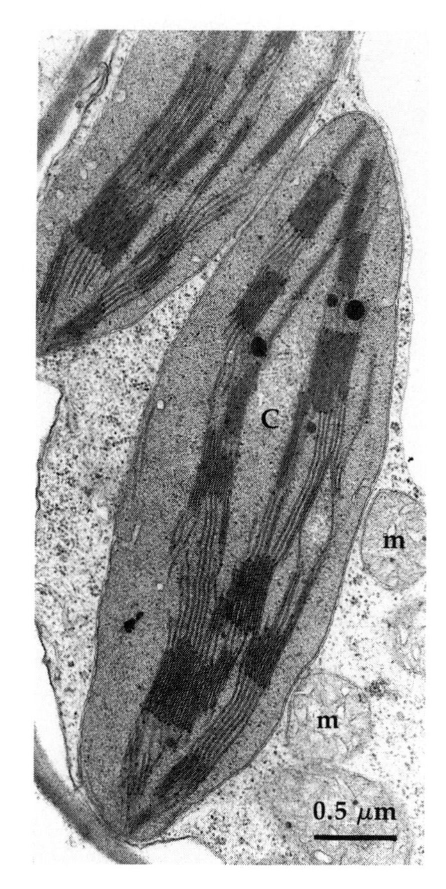

Figure 1.3 Example of a transmission electron micrograph. This TEM of an ultrathin section through a plant cell reveals the internal cellular detail of the chloroplast (C). Mitochondria (m) are also visible.

tween a TEM and a light microscope may be useful in beginning to understand the kind of data that can be produced using a TEM, it cannot be used to explain the details of image formation. Image formation in the TEM is fundamentally different from that of a light microscope and is described in detail in Chapter 4.

The SEM can be compared to a hand lens or a dissecting light microscope. All three of these instruments image surface features of bulk samples, providing a three-dimensional perspective, although the SEM, with its larger depth of field and greater resolution, has a distinct advantage (Figure 1.4). Since the electron beam (SEM) or light beam (hand lens or dissecting light microscope) need not pass through the specimen to obtain an image, the specimen can be thick. As with the TEM, the analogy between a dissecting light microscope and an SEM must be interpreted quite narrowly. The process of image formation is fundamentally different between the two instruments. Chapter 5 describes image formation in the SEM.

Each electron microscope has its own method of producing images and, as a consequence, a distinct set of advantages and disadvantages. In most cases, the advantages and disadvantages for each are balanced as a series of trade-offs, some of which microscopists can control and others of which they cannot. The specific combination of trade-offs depends on the instrument itself and on the particular set of experimental conditions imposed by specimen type and/or research protocols. To achieve the optimal balance, it is necessary to understand the function of each component of an EM and how the components work together to form an image. With this knowledge the microscopist not only can choose which of the two instruments (or both) will be the most useful for a particular application but also can balance the instrument variables that affect image formation.

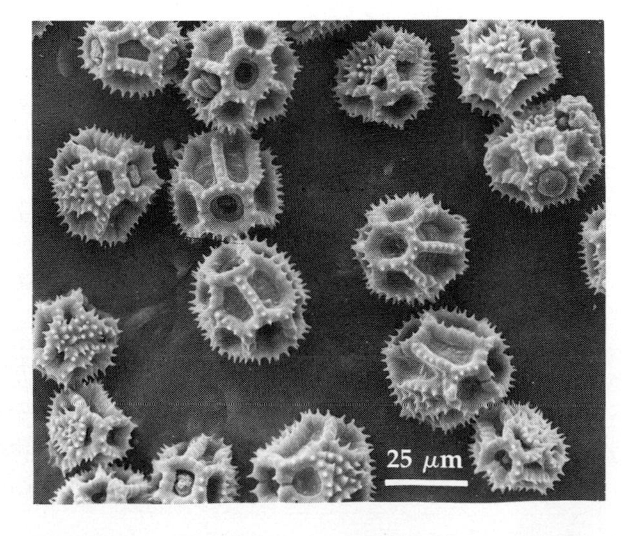

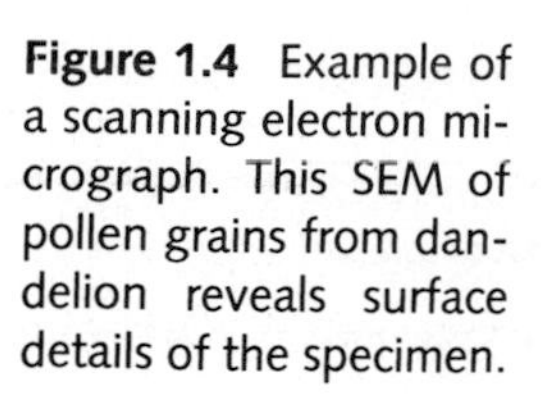
Figure 1.4 Example of a scanning electron micrograph. This SEM of pollen grains from dandelion reveals surface details of the specimen.

Specimen preparation for electron microscopy can be complex and often entails trade-offs that require careful consideration before a protocol is selected. Each of the following chapters is devoted, in part, to the task of explaining the individual instruments or the details of specimen preparation and achieving the optimal balance of trade-offs in each. Another important consideration for optimizing data obtained from the TEM or SEM is how they are presented. Chapter 9 explains the process of photography in the electron microscopes and the further processing of images using proper darkroom techniques to yield accurate and useful data (i.e., electron micrographs).

Since electron microscopes use electrons instead of light to carry the information generated during image formation, all images are black and white. Color itself is a function of visible light, and no visible light is used to generate images in an electron microscope. Color images may be generated using computerized imaging equipment on color monitors, but they are not indicative of any actual color in the specimen itself. Consequently, the usual dyes and stains used in light microscopy do not impart useful information in an EM image, unless they also happen to contain heavy metals that can contribute to electron image formation.

The annual proceedings of the Microscopy Society of America (before 1993 known as the Electron Microscopy Society of America) contain current information on improved and new electron microscopes and auxiliary equipment, as well as new methods of specimen preparation. Many new uses and modifications of electron microscopes and other microscopic imaging technology, along with new technology and applications of electron microscopy are also described in the proceedings each year. The proceedings are an excellent source of current information to supplement books, reviews, and specific disciplinary journal articles.

Other Microscopic Imaging Devices Provide Useful Data Correlative to Standard TEM and SEM Data

Although the specific subjects of this text are TEM and SEM, for many areas of study it is important to recognize the correlative value of light microscopy with both kinds of electron microscopy. The light microscope can provide images of surfaces using a stereoscope or dissecting microscope or images of internal structures of thick sections using a compound microscope. Use of specific colored dyes and/or fluorescent markers provides additional anatomical and analytical data. Specimen preparation is, in general, simpler than that for EM and therefore often provides valuable initial information that can be applied to designing studies done with equipment capable of higher resolution.

Another instrument that produces useful correlative data for electron microscopy is the laser scanning confocal microscope (LSM). The LSM uses a laser as an illuminating source to scan across a specimen. In the confocal mode, the laser can produce an image of a specific plane of defined depth of the sample, thus filtering out all other information above and below that plane. This capability is called optical sectioning or Z sectioning. The result is an image that is clearer, with more detail than a conventional light micrograph. The resolving power of an LSM is between that of a light microscope and an electron microscope, approximately 100 nm. (For more information on the LSM, see Pawley (1989) in Further Reading. Used in parallel, the light microscope, the laser scanning confocal microscope, the TEM, and the SEM provide unique yet complementary data. Scanning-probe microscopy (see Chapter 5 and Burleigh (1990) in Further Reading) and other near-field microscope methods (see Betzig and Trautman (1992) in Further Reading) are relatively new technologies for imaging specimens and constitute additional tools for data collection.

Both the TEM and SEM can be modified, resulting in instruments whose names aptly describe their functions. For example, the scanning transmission electron microscope (STEM) produces a transmitted image (as a TEM does) but uses a scanning beam (as an SEM does) (see Chapter 5). The analytical electron microscope (AEM) is a TEM, an SEM, or an STEM to which one or more analytical devices has been attached. In addition to producing the standard TEM, SEM, or STEM images, an AEM can also provide corroborative analytical data. An X-ray microanalyzer, for example, can provide qualitative and quantitative elemental composition and compositional mapping of a specimen at the subcellular level (see Chapter 8).

Another class of EM instruments are the scanning probe microscopes (SPM). Although these are not electron microscopes, they do produce high-resolution images of specimen surfaces. The scanning tunneling microscope (STM) and the atomic force microscope (AFM) are two examples of SPMs (see Chapter 5). The STM produces an image using the variations in tunneling current generated from a fine point of metal (the probe), which scans very close to the specimen surface. The STM can be used to image the surface of a conductive specimen at atomic level resolution. The atomic force microscope (AFM) is another scanning probe instrument. This instrument relies on the attractive forces that are generated when the electron clouds of two atoms are in close proximity. The AFM can be used to image nonconductive samples at near-atomic resolution.

The Evolution of Electron Microscopes Depended on Many Developments

With the first use of a simple light microscope in 1673, Antoni van Leeuwenhoek was able to produce magnified images of samples too small to be seen with the naked eye. In 1677 Robert Hooke built the first compound light microscope. Modifications of the light microscope continue to be made today, although the limits of resolution are imposed primarily by the use of light as the image-forming source.

In the 1870s Ernst Abbe developed a simple mathematical equation to define resolving power by relating the wavelength of light, the size of the lens aperture, and the nature of the medium through which the light passes. His equation was based on work done by Sir George Airy and Baron Rayleigh, who earlier had examined other optical phenomena.

In 1897 Sir J. J. Thomson discovered the electron. This work and his subsequent research, for which he received a Nobel Prize in Physics in 1906, demonstrated that the electron is a particle of very small mass, has a negative charge, and can be deflected by electric and magnetic fields.

In 1924 Prince de Broglie (1929 Nobel Prize in Physics) defined the wave nature and wavelength of electrons, although he did not use this information to postulate the possibility of an electron microscope. In 1926 H. Busch published his research on the movement of electrons in magnetic fields, which demonstrated that a magnetic field could act as a lens to focus electrons.

During the 1930s and 1940s, Walter Glaser worked on theoretical considerations in electron optics, including imaging properties of objective lenses and lens aberrations, and recognized the increased importance of spherical aberration in electron optics as compared to light optics. His work culminated in a classical volume entitled *Grundlagen der Elektronenoptik* (*Fundamentals of Electron Optics*) in 1952.

The development of the transmission electron microscope (TEM) in the early 1930s is credited to Max Knoll and Ernst Ruska at the High Voltage Laboratory at the Technological University in Berlin. As part of his student thesis and later his Ph.D. dissertation, Ruska built the first electromagnetic lens suitable for use in an electron microscope, and in 1933 he constructed the first true TEM, including lenses with pole pieces. With Bodo von Borries in 1939, Ruska built the first commercial TEM in the Siemens Company factory in Germany. Interestingly, in 1931 Reinhold Rudenberg had filed the first German patent application for the electron microscope on behalf of the Siemens Company, although he did not actually participate in its development (see Ruska (1986) in Further Reading).

In 1935 Knoll demonstrated the theory of the scanning electron microscope (SEM), based on the first theoretical considerations proposed by Sintzing. In 1938 von Ardenne built the first scanning electron beam instrument, which was essentially a STEM. The first SEM to resemble the present-day version was developed at the Radio Corporation of America by Vladimir Zworykin, James Hillier, and Gerald Snyder in 1942. Oatley, McMullan, and Smith at Cambridge University made vast improvements in SEM design between the years 1948 and 1952. T. E. Everhart and R. Thornley developed an efficient secondary-electron detector in 1960; and the first commercial SEM was introduced to the market by Cambridge Instruments in 1964.

Other developments related to electron microscopy include the design of an electron microprobe for localization of elements based on X-ray microanalysis in 1951 by Castaing (in his doctoral dissertation); the first semiconductor X-ray detector by Fitzgerald, Keil, and Heinrich in 1968; the scanning transmission electron microscope (STEM) in 1971 by Crewe and Wall; and the scanning tunneling microscope by Binnig and Rohrer in 1985.

Although this brief history recounts only the development of electron microscopes, another important area of research in the field has been the development of various protocols to prepare specimens for examination. This research is especially crucial for biological applications of both TEM and SEM, since most biological samples are predominantly water, which, with few exceptions, is incompatible with the high vacuum required by most electron microscopes. Protocols that preserve specimens in as lifelike a state as possible and that minimize or eliminate artifacts are crucial for the collection of useful ultrastructural data. The most common procedures for preparing most biological samples and some materials-science samples are described in Chapters 6 and 7. Methods for specific specimens and experimental designs can be found in the disciplinary journals.

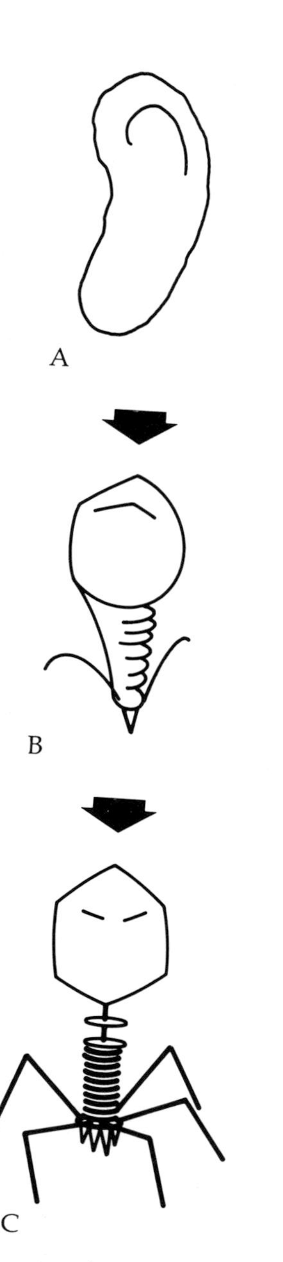

Figure 1.5 Resolving power. As resolving power increases, so does the ability to distinguish the fine detail of the specimen. Poor resolving power (A) does not allow structural detail to be discerned. As resolving power increases, more details become visible (B and C).

The Key to the Usefulness of Electron Microscopes Is Their Resolving Power

As stated earlier, resolving power is the ability to distinguish between two closely spaced points (Figure 1.5). By definition, the smaller the distance that can be distinguished between two points, the greater the resolving power. The resolving power of both light microscopes and TEMs is dependent on and limited by the wavelength of the energy source used to form the image, as well as other considerations that are explained in detail in Chapter 4.

In brief, for a TEM resolving power is related to the wavelength and numerical aperture of a lens by an equation derived by Ernst Abbe. Abbe's equation states: resolving power = 0.61λ N.A., where λ is wavelength, and N.A. is the numerical aperture of the lens. Wavelength is defined as the distance from one wavecrest to the next. Numerical aperture is defined as $n\sin\alpha$, where n is the refractive index of the medium of the lens (approximately 1 for the vacuum in a TEM), and α is one-half the acceptance angle of the lens (see Chapter 4).

Although the relationship between resolving power and wavelength was defined for light-emitting points, it does have application in defining resolving power for the TEM. An approximation for the light microscope is that resolving power = 0.5λ. Information is certainly present in an image, but details of structures cannot be resolved if they are smaller than one-half the wavelength of the imaging source used.

In the case of a light microscope (λ = 400 nm to 700 nm) with a resolving power of approximately 200 nm, bacterial flagella, DNA, most small viruses, and cell membrane structures such as endoplasmic reticulum cannot be visualized. These can be seen only with a resolving power of 50 nm or better, which is easily obtainable in a TEM because of the smaller wavelength of electrons. Other factors affecting the resolution of a TEM are discussed in Chapter 4.

The resolving power of an SEM depends on factors somewhat different from those that affect the resolving power of a TEM. The TEM operates in a flood-beam manner: A large portion of the sample is illuminated with a beam of electrons. Electrons that pass through the sample are used to produce a transmitted, optical image. In the SEM, on the other hand, the beam of electrons is focused to a very narrow point on the sample surface and scans across the sample. The interaction of the beam with the sample produces secondary electrons that are used to generate the image. The image is not produced in an optical manner, as is the TEM image. For an SEM, the diameter of the beam is the ultimate limiting factor affecting resolution; no features whose spacing is less than the diameter of the beam can be resolved. Chapter 5 discusses SEM image formation in more detail.

The advantages of resolving power and magnification are often incorrectly interchanged. High-magnification capability is frequently cited as the primary reason for using an electron microscope, but it is increased resolving power that provides the advantage. It is useful, however, to consider resolving power and magnification together. Assuming that the necessary resolving power exists, the microscopist can choose the appropriate magnification to distinguish the desired features of a specimen (Table 1.2). Smaller detail requires greater magnification to be visible to the human eye.

Units of Length Used in Electron Microscopy

Several units of length are common in electron microscopy. These units are based on the Système Internationale des Unites (SI), which was devised in 1960 to provide standardized units of measurement. The

TABLE 1.2
Comparison of the Resolving Power and the Minimum Magnification Required to Image Selected Specimens

SPECIMEN	RESOLVING POWER REQUIRED	MINIMUM MAGNIFICATION RANGE REQUIRED
Bacteria	1–2 µm	1,000–1,500×
Viruses	20–60 nm	20,000–80,000×
Large atoms	0.1 nm	2,000,000×

most widely used units of length are the micrometer (commonly used for SEM micrographs) and the nanometer (commonly used for higher-magnification TEM micrographs). The relationships for these units are as follows: 1 mm = 1,000 μm (10^{-3} mm), and 1 μm = 1,000 nm (10^{-6} mm). The angstrom (Å; 1 nm = 10 Å) is used earlier scientific literature and occasionally in modern literature. In modern SI terminology the angstrom is referred to as a decinanometer (dnm).

Safety in the EM Lab Is an Important Consideration

The protocols for preparing specimens and procedures for operating equipment vary among laboratories. The specific safety requirements for electron microscopy at each educational institution, company, or government laboratory should be understood and followed. Safety recommendations also appear throughout this text, especially for the specimen preparation procedures, although specific guidelines must be provided by laboratory instructors and EM facility directors. A set of references is included in this chapter following the Further Reading section.

FURTHER READING

Agar, A. W., R. H. Alderson, and D. Chescoe. 1974. *Principles and Practice of Electron Microscope Operation.* North-Holland, Amsterdam, and Elsevier, New York.

Betzig, E., and J. K. Trautman. 1992. Near-field optics: Microscopy, spectroscopy, and surface modification beyond the diffraction limit. *Science* 257:189–195.

Burleigh Instruments, Inc. 1990. *The Scanning-Probe Microscope Book.* Burleigh Instruments, Fishers, NY.

Cosslett, V. E. 1951. *Practical Electron Microscopy.* Academic Press, New York. (Cosslett's rule)

Hawkes, P. W., ed. 1985. *The Beginnings of Electron Microscopy.* Academic Press, New York.

Holwill, M. E., and N. R. Silvester. 1973. *Introduction to Biological Physics.* Wiley, New York.

Knoll, M., and E. Ruska. 1932. Das Elektronenmikroskop. *Z. Physik* 78:318–339.

Meek, G. A. 1976. *Practical Electron Microscopy for Biologists.* Wiley, New York.

Newberry, S. P. 1992. *EMSA and Its People: The First Fifty Years.* Electron Microscopy Society of America, Boston.

Pawley, J., ed. 1989. *The Handbook of Biological Confocal Microscopy.* IMR Press, University of Wisconsin, Madison.

Ruska, E. 1934. Über Fortschritte im Bau und in der Leistung des magnetischen Elektronenmikroskops. *Z. Physik* 87:580–602. (Original description of the TEM)

Ruska, E. 1986. The emergence of the electron microscope: Connection between realization and first patent application. Documents of an invention. *J. Ultrastruct. Mol. Struct. Res.* 95:3–28.

Weber, R. L. 1988. *Pioneers of Science: Nobel Prize Winners in Physics.* Second edition. Hilger, Philadelphia. (Scientists involved in the development of EMs)

Wischnitzer, S. 1981. *Introduction to Electron Microscopy.* Second edition. Pergamon, New York.

LABORATORY SAFETY: FURTHER READING

Alderson, R. H. 1975. *Design of the Electron Microscopy Laboratory: Hazards and Safety Precautions.* American Elsevier, New York.

Barber, V. C., and D. L. Clayton, eds. 1985. *Electron Microscopy Safety Handbook.* EMSA, San Francisco Press, San Francisco.

Bastacky, J., and T. L. Hayes. 1985. Safety in the scanning electron microscope laboratory. *Scanning* 7:255–272.

Mackinson, F. W., and R. S. Stricoff, L. J. Partridge, Jr., and A. D. Little, eds. 1980. *Pocket Guide to Chemical Hazards.* National Institute for Occupational Safety and Health/Occupational Safety and Health Administration, Washington, DC.

Parsons, D. F., V. A. Phillips, and J. S. Lally. 1973. *Handbook of X-Ray Safety for Electron Microscopists.* Radiation Committee, Electron Microscopy Society of America, San Francisco.

Smithwick, E. B. 1985. Cautions, common sense and rationale for the electron microscopy laboratory. *J. Electron Microsc. Technol.* 2:193–200.

2

Electron Sources and Electron Lenses

ELECTRON SOURCES

The first stage in the process of creating images in any electron microscope is the production of a beam of electrons. The electrons are generated in the electron gun, which consists of the anode and the Wehnelt assembly (Figure 2.1). The electron gun is located in the gun chamber, which is usually found at the top of the column (Figure 2.2A). The column is a cylindrical structure that houses, in addition to the electron gun, the electromagnetic lenses and an area for the samples. Three major types of electron gun are used in electron microscopes: tungsten-hairpin filament guns, lanthanum-hexaboride guns, and field-emission guns.

Tungsten-Hairpin Guns Are Most Common

Tungsten-hairpin guns, used in most electron microscopes, are considered the standard. The source of electrons is a loop of tungsten

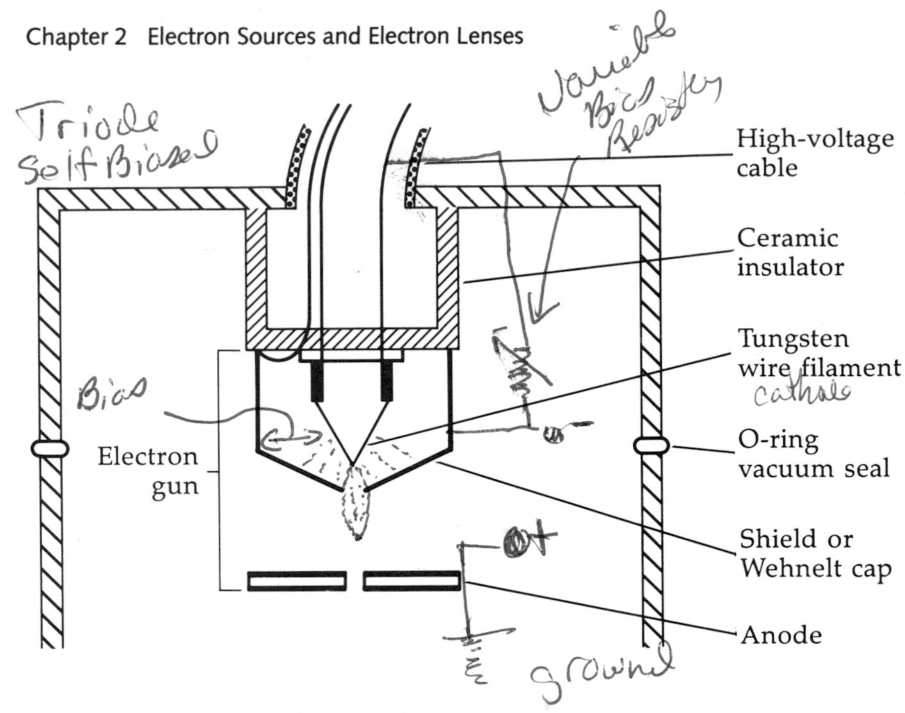

Figure 2.1 Schematic of a typical electron gun chamber and associated components.

wire that is bent into a hairpin shape (thus the name; Figure 2.2B; see also Figures 2.1 and 2.4). The loop of tungsten wire is called the filament, and it functions as a cathode. During operation, a small voltage differential applied across the terminals of the filament causes heating of the wire because of electrical resistance. The flow of current through the wire is called filament current. Through a process called thermionic emission, a cloud of electrons is produced at the tip of the filament. Because the filament is heated, this gun is called a hot-cathode gun. Although a small voltage differential is maintained between the filament wires, the entire filament is at a very high negative potential compared with the anode, which is at ground or zero potential. The anode is, thus, positive with respect to the filament and forms a powerful attractive force for the negative electrons emitted from the filament tip. The electrons are accelerated toward the anode. Some strike the anode and do not contribute to image formation. Many pass through

A B

Figure 2.2 Electron gun chamber and the Wehnelt assembly. (A) A typical electron gun chamber with the top portion tilted backward. A gun chamber normally is opened to change a burned-out filament. (B) The Wehnelt assembly with the Wehnelt cap removed, showing the tungsten-hairpin filament.

the anode aperture and produce the beam of electrons. The beam of electrons is directed down the column, where it is condensed and focused by the electromagnetic lenses.

The Wehnelt cap or shield is a metal cover that surrounds the filament (see Figures 2.1 and 2.2). During operation, the shield is maintained at a potential several hundred volts more negative than the filament. The difference in voltage between the filament and shield is called the bias voltage. The Wehnelt cap with the negative bias has a repulsive effect on the electrons and acts as an electrostatic lens, bringing the electrons to a focus or crossover just in front of the anode.

The voltage difference between the filament and anode is called the accelerating voltage. It determines the energy and wavelength of the electrons as they move down the column. Accelerating voltages for most TEMs range from about 60,000 V to 400,000 V, although some special high-voltage instruments operate at one million volts or more. Most SEMs are used at accelerating voltages of 5,000 V to 30,000 V, with increasing emphasis being placed on SEMs that are able to operate at accelerating voltages as low as several hundred volts.

The application of voltage across the terminals of the filament is a very critical process called filament saturation. It is not possible to select a predetermined voltage. Electron microscopes have a control device, usually labeled "filament" or "emission," for slowly and carefully applying the voltage, which in turn causes the flow of filament current. A plot of beam current as a function of filament temperature shows an increase in beam current to a certain point, called the first peak, where it decreases briefly and then starts to rise again (Figure 2.3). Most filaments begin to produce an electron emission around some imperfection on the side of the loop, which causes the first peak. As the filament current increases, the point of maximum electron emission moves toward the tip of the filament. Eventually a point of saturation, at which the beam current ceases to rise with increased filament current, is reached. Increasing the filament current beyond this point is called oversaturation and greatly decreases the life of the filament: The filament may melt in a few minutes, compared to a normal life of many hours (Figure 2.4).

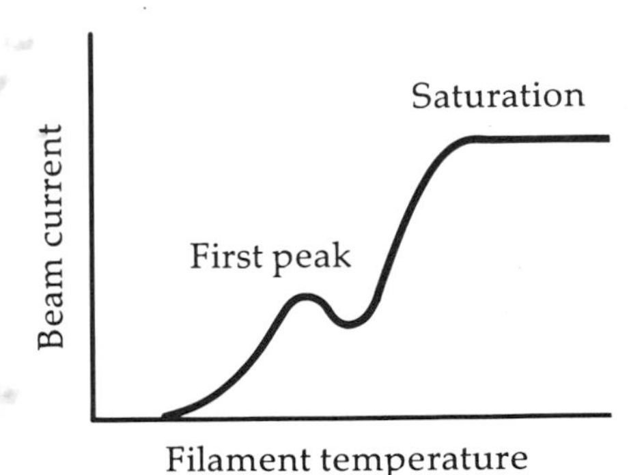

Figure 2.3 Tungsten-filament saturation curve. Most tungsten filaments cause two peaks in electron-beam current. The first peak may produce an electron emission around an imperfection on one side of the loop of tungsten. The emission giving rise to the second peak is from the tip of the wire loop.

With most electron microscopes saturation is monitored by observing a current meter; with the TEM it is also monitored by observing an optical image of the filament on the viewing screen (Figure 2.5A); and with the SEM it is also monitored by observing a display on a cathode-ray tube called a linescan (Figure 2.5B).

Saturation occurs because of the method used to apply the bias voltage to the Wehnelt cap. The entire arrangement is called a self-biased gun. A variable-bias resistor is inserted between the Wehnelt cap and the filament. As soon as electron emission begins from the filament, current also flows from the Wehnelt cap to the filament through the

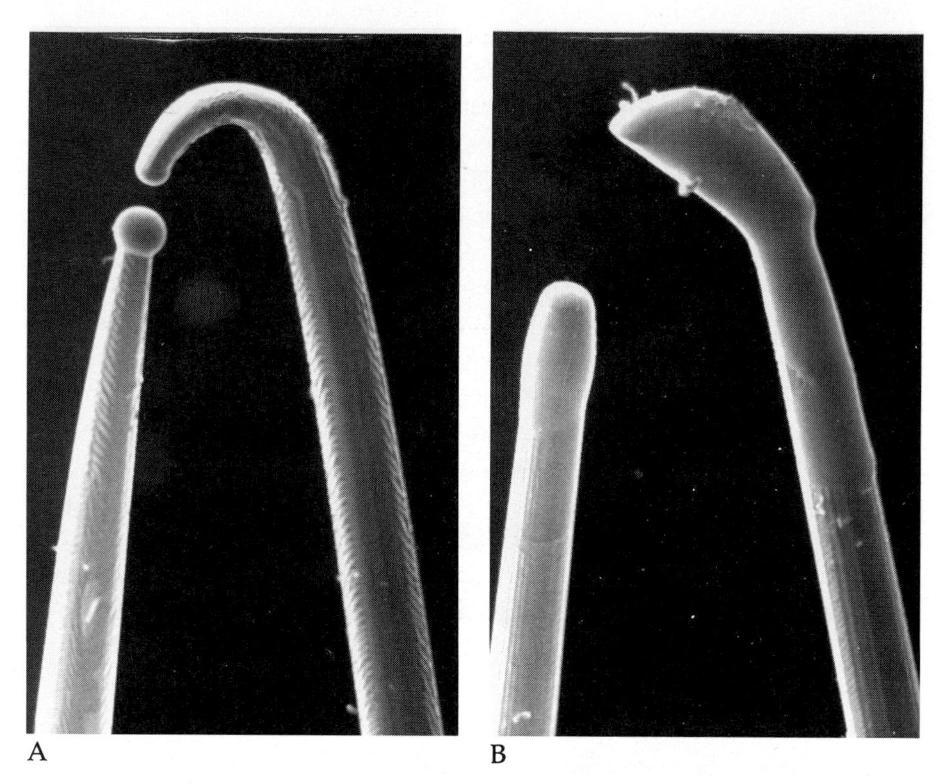

Figure 2.4 Burned-out tungsten filament. (A) A filament that burned out after many hours of proper saturation. The edge of the wire has decreased in diameter because of evaporation. Eventually the wire melted. (B) A filament that burned out after one hour of oversaturation. The wire shows little thinning, and there is a large accumulation of molten tungsten at the end.

resistor. The flow of current through the resistor causes the Wehnelt cap to develop the negative potential or bias that begins to hinder further emission of electrons from the filament because the negative bias repels the negative electrons. Eventually, the resulting bias voltage becomes sufficient to stop further emission of electrons; this is the point of saturation.

The amount of bias voltage developed through the resistor can be controlled by a knob that changes the value of the resistance. Selecting

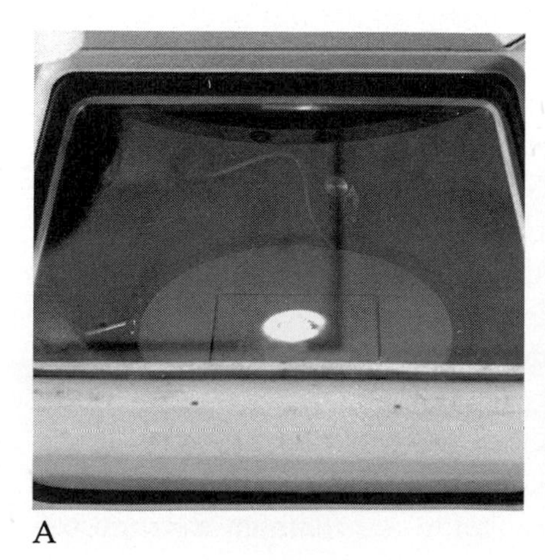

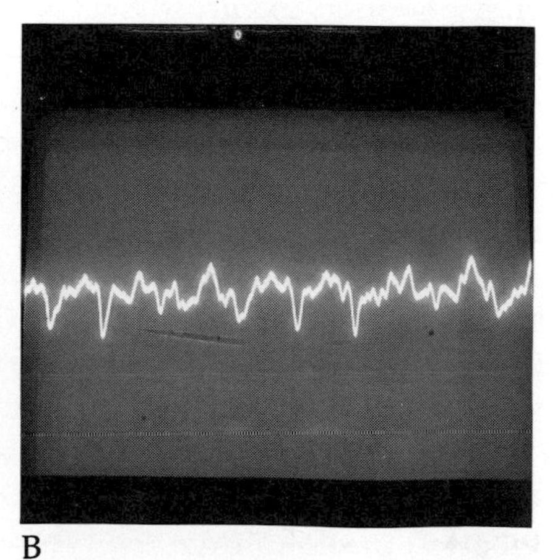

Figure 2.5 Monitoring filament saturation. (A) Filament saturation in a TEM normally is monitored by watching the image of the filament on the phosphor-coated viewing screen. (B) Filament saturation in an SEM normally is monitored by viewing the linescan on a cathode-ray tube.

the proper amount of bias is very important because it has a direct effect on the electron-beam crossover that is formed between the Wehnelt cap and the anode. If too little bias is developed, the crossover is formed poorly, and the beam current is low. If too much bias is developed, there is a strong repulsive effect from the Wehnelt cap, many electrons are repelled to the filament, and the beam current is also low. Low beam currents are not desirable because they impede the performance of the electron microscope.

The tungsten filament is the most commonly used type of electron gun because it offers a number of advantages. It provides a relatively stable source of electrons, it is inexpensive to replace, and it does not require an ultrahigh vacuum for proper operation. Its useful life is usually between 50 and 150 hours. During operation, the tungsten metal is slowly vaporized from the filament. Eventually the wire becomes very thin and separates (see Figure 2.4A); the filament has burned out. When this happens, the entire electron gun assembly must be removed and carefully cleaned. A new filament is then inserted, and the centering of the filament and the spacing between the Wehnelt cap and the filament are carefully adjusted.

Lanthanum-Hexaboride Guns Offer Some Advantages

The second most common type of electron gun, the lanthanum-hexaboride gun, is commonly called a lab-six gun (after the chemical formula for lanthanum hexaboride, LaB_6). The LaB_6 gun is very similar to a tungsten gun. A crystal of LaB_6 is mounted on a rhenium or tungsten wire support housed in a Wehnelt assembly (Figure 2.6). A voltage applied between the two terminals of the wire causes heating of the crystal. The hot crystal then emits electrons.

A LaB_6 emitter offers three major advantages over a tungsten-hairpin gun. First, the beam current obtained under equivalent conditions can be up to ten times that of a tungsten gun. The amount of beam current is often referred to as beam brightness because this is how the effect is observed in the electron microscope. The LaB_6 gun is thus brighter than a tungsten-hairpin gun. Second, the electrons are emitted from a more narrow point in the lanthanum-hexaboride gun, and thus a crossover of smaller diameter is produced between the gun and the anode. These two factors usually enable an SEM to give superior resolution (see Chapter 5). Third, the energy spread of the emitted electrons is about one-half that of the tungsten-filament gun, a factor that enables both SEMs and TEMs to provide improved resolution. A fourth advantage is that of a longer useful life, often ten times that of a tungsten gun.

Figure 2.6 Lanthanum-hexaboride electron source. A crystal of lanthanum hexaboride is held between tungsten or rhenium wire supports. The flow of current through the wire heats the crystal, which then emits electrons.

There are, however, three disadvantages with the LaB_6 gun. First, a better vacuum is required for proper operation than is needed for a tungsten gun, although most modern high-performance instruments have sufficient vacuum. Second, saturation is usually far more critical and time-consuming than it is with a tungsten gun. Reaching saturation takes only a minute or so with a tungsten gun, but may take a half-hour or more with a LaB_6 gun because the crystal must be warmed up very slowly. Finally, the LaB_6 gun is far more expensive to replace than a tungsten filament is. Costs vary depending on supplier, but LaB_6 emitters can cost ten to one hundred times as much as tungsten filaments do.

Field-Emission Guns Give the Best Resolution

The third most common type of electron gun, the field-emission gun, is found at present in only a small percentage of all instruments, although it may become more common as designs improve. The construction of a typical field-emission gun and electron emitter is shown in Figure 2.7. The operation of a field-emission gun relies on a principle of physics called field emission, in which, in a high vacuum, electrons are physically drawn off a very finely curved tungsten tip with an applied voltage field. The accelerating voltage for such a gun is the voltage between the tungsten crystal and the second anode. A positive voltage of up to 2,000 V on the first anode controls the amount of electron emission. Nothing is heated with this type of gun; thus, it is called a cold-cathode gun, as opposed to the tungsten-hairpin and LaB_6 types, which are hot-cathode guns.

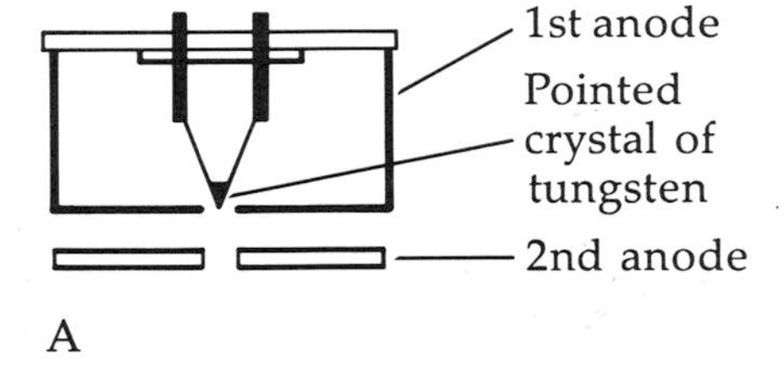

Figure 2.7 Field-emission gun. (A) A crystal of tungsten formed to a very narrow point is mounted on a loop of tungsten wire. The surrounding cap resembles a Wehnelt cap but functions as a first anode because it is positive with respect to the tungsten crystal. (B) The crystal of tungsten.

A field-emission gun offers four advantages over a tungsten or lanthanum-hexaboride gun. First, the brightness may be up to a thousand times greater than that of a tungsten gun. Second, the electrons are emitted from a point more narrow than that in the other sources, and thus a crossover of smaller diameter is produced between the gun and the anode. These two factors enable an SEM equipped with a field-emission gun to give far superior resolution than that of an SEM equipped with either a tungsten gun or a LaB_6 gun. Third, the energy spread of the emitted electrons is about one-tenth that of the tungsten-hairpin gun and one-fifth that of the LaB_6 gun, which enables both SEMs and TEMs to give superior resolution. Finally, the field-emission gun has a very long life, up to a hundred times that of a tungsten gun.

The field-emission gun has three disadvantages. First, the microscope must be designed for this type of gun from the beginning. The gun cannot simply be plugged in as a replacement for one of the other types because the gun chamber must be designed for the double-anode arrangement. Second, the vacuum levels necessary for proper operation are several orders of magnitude greater than those required for the other guns. The minimum vacuum needed is usually 1×10^{-8} Pa (1×10^{-10} torr), but most electron microscopes with a standard tungsten gun routinely achieve a vacuum of only 1×10^{-4} Pa (1×10^{-6} torr) in the gun chamber. Designing an electron microscope that can achieve this vacuum level consistently is complex, and the resulting instruments are therefore proportionately much more expensive than their conventional counterparts. The third disadvantage of the field-emission gun is that it tends to produce a beam of electrons that is not stable in current—i.e., the quantity of electrons may change from one minute to the next. This may be a major disadvantage in certain quantitative analytical electron microscopy procedures.

ELECTRON LENSES

After the beam of electrons has been formed and has passed through the anode aperture, it is controlled by the lenses. In the early days of electron microscopy, some machines used electrostatic lenses. With these, metal plates that carried an electrical charge were used to deflect the beam of electrons. This method did not work well and was abandoned. There is an electrostatic focusing effect from the Wehnelt cap in modern electron microscopes, but this is not considered to be a true lens. The lenses in all electron microscopes are magnets (or, as they are more properly called, electromagnets, because their magnetic properties depend on the passage of an electrical current).

Design

If a nail is wrapped with insulated wire and the ends of the wire are connected to a battery, an electromagnet with magnetic lines of force is created. The same principle is applied in an electromagnetic lens. A simple electromagnetic lens can be made by winding a wire around a hollow core and covering it with a special soft iron covering called a shroud (Figure 2.8). The beam of electrons passes through the opening in the center and is focused and deflected by the magnetic lines of force. This primitive type of lens was used in early electron microscopes. To construct more powerful microscopes, stronger lenses had to be produced. Stronger lenses were constructed by wrapping the shroud around and drawing it in at the center leaving only a very narrow gap without the covering. This design concentrates all the magnetic lines of force in the gap, creating a very powerful lens. Such a lens is called a pole-piece lens (after the poles of a magnet). The pole pieces are machined to very precise tolerances and must be handled and cleaned with extreme caution because even a hairline scratch can distort their magnetic field so much that they have to be replaced. Many lenses, especially those used in TEMs, have to be cooled with water to prevent overheating.

All electromagnetic lenses are controlled or focused by varying their current, usually with a control device on the microscope itself. Allowing more current to flow results in a shorter focal length, thereby

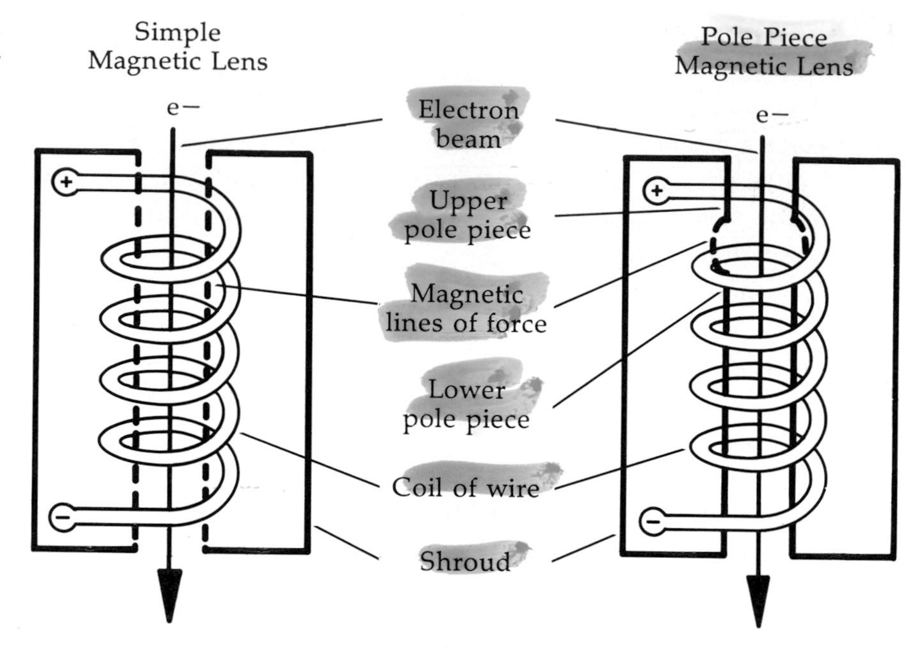

Figure 2.8 Magnetic lenses. A simple magnetic lens has a soft iron shell (or shroud) that is C-shaped on each side in cross section. This design does not produce a powerful lens. A pole-piece magnetic lens has the shroud drawn inward to leave only a very narrow gap. The magnetic lines of force are concentrated within this gap, thus producing a powerful lens.

increasing the power or strength of the lens. An electromagnetic lens performs functions similar to those of a glass lens in a light microscope: As a condenser lens, it can condense the beam of electrons; as an objective lens, it can focus on the specimen; as a projector lens, it can project an image onto a screen. Specific functions are discussed in more detail in the chapters on each instrument.

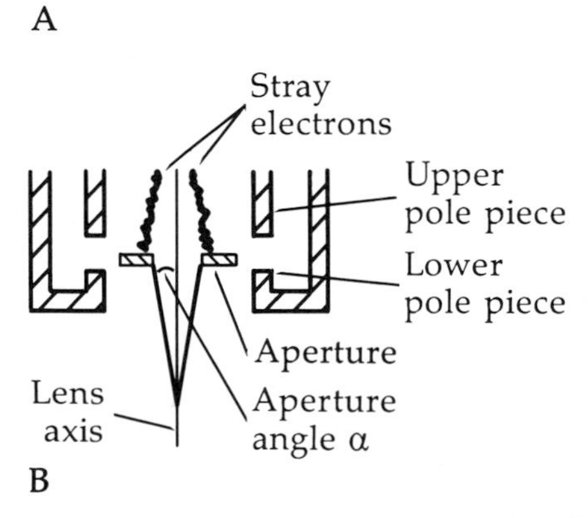

Figure 2.9 Electron microscope apertures. (A) The apertures used in electron microscopes come in many shapes and sizes. Some designs have three different sizes on a thin strip of metal. (B) Apertures shape the beam of electrons, remove stray electrons, and have many secondary effects specific to their location in the electron microscope.

Apertures

Apertures are used in conjunction with the electromagnetic lenses in electron microscopes. An aperture is a piece of metal with a small round hole, typically 30 μm to 1,000 μm in diameter, depending on its placement and specific function (Figure 2.9A). The primary use of apertures is to limit the diameter of the beam of electrons or to eliminate stray or widely scattered electrons (Figure 2.9B). Apertures have many secondary effects of major importance in electron microscope operation. These effects are discussed in Chapters 4 and 5.

Defects

A number of defects are associated with electromagnetic lenses, just as glass lenses in light microscopes have defects. These defects limit the resolution from electron microscopes, and some require careful and constant attention from the microscopist.

One lens defect is called astigmatism, which is defined as a defect of lens symmetry that results in differing lens strengths in two directions at right angles—i.e., the field of pull of the lens is not uniform. This is the same type of astigmatism that many individuals have in their vision. In electromagnetic lenses, astigmatism is sometimes the result of manufacturing defects and more commonly the result of contamination. It is not possible to make a perfect lens, even though manufacturers go to great lengths to make the lenses as uniform as possible. In addition, small amounts of contamination enter the column from the oils used in the pumps and from the specimens. As these materials deposit on the pole pieces or on apertures, they build up an electrical charge that deflects the beam of electrons slightly and therefore alters the force field of the lens. Such a change in the force field then alters the beam of electrons so that in cross section it is no longer round, but oval. The distorted beam of electrons results in distorted images with much less resolution. Fortunately, astigmatism is easy to correct in most circumstances. Manufacturers add stigmator coils, called octapole stigmators, which are associated with one or more of the lenses. Eight small electromagnets are placed around the column. The current in these coils may be varied by controls on the microscope. The coils are wired so that the electromagnetic field from them may be shifted around the cross-sectional axis of the column. Adjusting the controls forces the distorted beam of electrons back into a round shape.

Another lens defect is that of spherical aberration. With this defect, the peripheral electrons are deflected more than the electrons closer to the center are (Figure 2.10). Spherical aberration can be described by the equation $d_s = \frac{1}{2}C_s\alpha^3$, where C_s (the spherical aberration coefficient) is directly related to the accelerating voltage and the focal length of the lens and α is the aperture angle. The aperture angle is the angle formed by the edge of the electron beam as it exits the lens aperture and the center of the crossover (see Figure 2.9B). Unfortunately, it is not possible to eliminate spherical aberration completely, and it is one of the major factors that limits the practical resolution of electron microscopes to a level well below what theoretically is possible. Reducing the focal length of a lens, the diameter of the aperture associated with the lens, and the accelerating voltage are the most useful means of the effect of spherical aberration. These means of reducing spherical aberration, however, involve trade-offs that are discussed in Chapters 4 and 5.

Another lens defect is chromatic aberration (the name is derived from the Greek word *chroma,* meaning "color" or "wavelength"). It is the result of electrons in the beam having different velocities and thus different wavelengths. The differing wavelengths cause the electrons to be brought to focus at different points, which blurs the image. Chromatic aberration can be defined by the equation $d_c = (\Delta E/E)C_c\alpha$, where $\Delta E/E$ is the fractional variation in electron-beam voltage, C_c is the chromatic aberration coefficient, which is related to the focal length of the lens, and α is the aperture angle. There are several methods of reducing chromatic aberration. One is to reduce the variation in the accelerating voltage. (Most modern electron microscopes have excellent

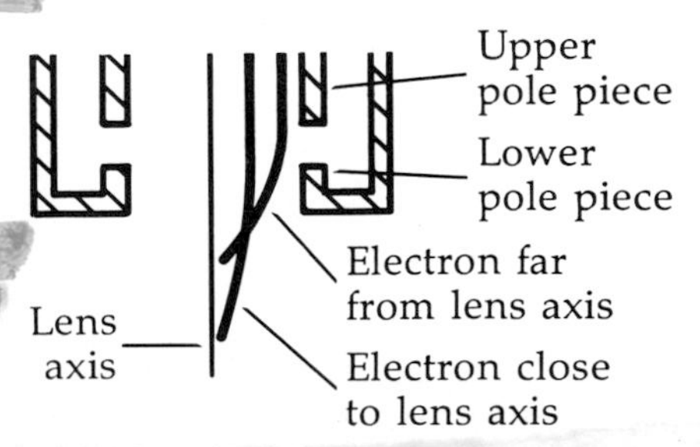

Figure 2.10 Spherical aberration in a lens causes the peripheral electrons to be deflected more than the electrons that are closer to the center. It is one of the major factors limiting the practical resolution of electron microscopes.

regulation of power supply, so the contribution from accelerating voltage variation to chromatic aberration is minimal.) A second method is to use an electron gun with a lower energy spread of the emitted electrons (i.e., a LaB_6 gun or a field-emission gun). Other methods include reducing the focal length of the lens, the angular aperture, or variation in the lens current.

FURTHER READING

Goldstein, J. I., D. E. Newbury, P. E. Echlin, D. C. Joy, C. Fiori, and E. Lifshin. 1981. *Scanning Electron Microscopy and X-Ray Microanalysis. A Text for Biologists, Materials Scientists, and Geologists.* Plenum, New York. (General reading)

Hohn, F. J. 1985. Development and use of high brightness lanthanum hexaboride electron guns. *Scanning Electron Microsc.* 1985(4):1327–1338. (Lanthanum-hexaboride electron guns)

Podbrdsky, J. 1986. High current density magnetic electron lenses in modern electron microscopes. *Scanning Electron Microsc.* 1986(3):887–896. (Magnetic lenses)

3

Vacuum Systems

Vacuum systems are critical to the operation of all electron microscopes and are a large portion of the microscope. They also play an important part in many aspects of specimen preparation. Because of the ubiquity of vacuum systems in the EM laboratory (and their fallibility), at least a descriptive understanding of them is necessary.

Electron Optics Requires a Vacuum

In the microscope itself, regardless of the type of instrument, a vacuum is necessary for several reasons. First, an electron beam cannot be generated or maintained in a gas-filled environment. If a gun chamber, where electrons are being accelerated by a multikilovolt potential, contained many gas molecules, some would become ionized. This would lead to random discharges (arcing), which would result in beam instability. A second major drawback to having gas within the gun chamber is the interaction between reactive gases (such as the oxygen from air) and the heated electron emitter. With the standard tungsten filament, even a small amount of oxygen would react with the filament wire at thermionic-emission temperatures, causing it to burn out. The highly efficient lanthanum-hexaboride (LaB_6) emitters used in some newer microscopes are more sensitive than the tungsten emitters (see Chapter 2). The LaB_6 emitter reacts with most elements (e.g., nitrogen gas from the air) when hot. Pressures in the gun chamber greater than about 10^{-4} Pa (a relatively high vacuum) cause the emitter to lose efficiency. Field-emission guns are even more labile to oxidation and require vacuums about a thousand times greater.

Even if it were possible to generate a suitable electron beam without a vacuum, beam transmission through the electron optic columns would be hindered by the presence of too many gas molecules. The beam-scattering interactions would lead to lower specimen image contrast. Thus, the vacuum must be sufficient to have a low probability of the electrons of the beam colliding with gas molecules (i.e., a long mean free path). A final consideration is the effect of gas molecules, some arising from the specimen or microscope themselves. If these gases remain in the column, they can form compounds that condense on the specimen, lowering contrast and obscuring detail. (A partial exception to the preceding discussion is the environmental SEM, described later in this chapter and in Chapter 5.)

Vacuum systems also play major roles in specimen preparation for electron microscopy. Sputter coaters, evaporators, plasma and ion mills, glow-discharge and freeze-etch specimen preparation equipment all require vacuum systems. The vacuum requirements and systems employed with some of these techniques are discussed under their separate headings.

The level of expertise that electron microscopists should have regarding vacuum systems depends on how much of the microscope they maintain and on how much specimen preparation is done. Good vacuum "common sense" is generally enough for running most modern EMs. A general knowledge of basic principles is helpful for sample preparation in vacuum devices or for equipment maintenance. It is toward these levels of knowledge that this chapter is directed. For design of or modifications to vacuum equipment, a considerably more rigorous and thorough treatment, beyond the scope of this text, is recommended.

What a Vacuum Is

Any meaningful discussion of vacuum systems requires the establishment of a descriptive knowledge of vacuum dynamics and terminology. The current SI standard unit for gas pressure (and hence vacuum) is the pascal (Pa), which corresponds to a force of 1 N/m^2 (N = newton). In the past, vacuum measurements often were presented in millibars (1 millibar = 1 mbar = 10^3 dynes/cm^2) or in terms of the height of a mercury column that they could support, with 1 mm of mercury (Hg) being equal to 1 torr (Table 3.1). Although the pascal is the standard unit of vacuum measure, torr and mbar units are still found in much of the classical literature and on older machines, making an understanding of both units worthwhile.

In addition to understanding the units used in vacuum measurement, it is helpful to have an idea of how gas behaves at lower pressures. Under standard atmospheric conditions, the air pressure at sea level supports a column 29.92 inches (760 mm) high in a mercury barometer. This level corresponds to a pressure of 101,325 Pa. At room temperature and pressure, a cubic meter of air contains about 2.5×10^{25} molecules which are in constant motion. Each molecule travels a mean distance (mean free path) of about 65 nm between collisions. As a typical EM vacuum is reached (approximately 10^{-4} Pa), the number of molecules is reduced by eight orders of magnitude, and the mean free path increases to about 6.5 m. Under these extremely different conditions, the gases that make up air behave in different ways.

Air is a mixture of gases, the condition and composition of which vary somewhat depending on season and location. Dry air is about 78% nitrogen, 20% oxygen, 0.033% carbon dioxide, and 0.934% argon, with the balance made up of other trace gases. In normal room air, the carbon dioxide level is somewhat elevated, and water vapor makes up 0.5% to 1% of the volume, making it the third most abundant gas. Within the vacuum system, the relative proportions of the gases change because they are pumped at different rates. Some additional gases arise from sample volatiles, and beam interaction products may also form high proportions of the residual gases.

Gas flows as a fluid at ambient pressures and in the vessel sizes used in EM. It follows the contours of the system containing it, as long as the flow rate is slow enough to avoid turbulence. Gas under these conditions moves by viscous (laminar) flow. Most collisions experienced by such gas molecules are with other gas molecules, and the mean free path is correspondingly short. This is the normal condition at the start of a pump-down cycle in EM laboratory equipment.

As the pressure decreases (vacuum increases) the frequency of collisions between particles decreases, and the interactions between the molecules and the vacuum system walls become increasingly important in the kinetic picture. When the mean free path of the molecules becomes larger than the diameter of the vessel, the gases are said to move by molecular flow.

Vacuum levels are often given qualitative values when discussed in EM applications. Table 3.2 illustrates some of these qualitative vacuum ranges and the type of gas flow that occurs at the different vacuum conditions.

TABLE 3.1
Common Vacuum Units

1 torr (T) = 1 mm Hg

1 micron, or micrometer (μm) = 10^{-3} torr

1 torr = $\frac{1}{760}$ atm

1 pascal (Pa) = 1 newton/m^2 = 7.5×10^{-3} torr

1 pascal = $\frac{1}{101,325}$ atm

1 millibar (mbar) = 10^3 dynes/cm^2 = 0.75 torr

1 millibar = $\frac{1}{1,013}$ atm

TABLE 3.2
Qualitative Vacuum Ranges

Laminar Flow Range	
Low or rough vacuum	1 atm–100 Pa
Medium or fine vacuum	100–0.1 Pa
Molecular Flow Range	
High vacuum	0.1–10^{-4} Pa
Very high vacuum	10^{-4}–10^{-7} Pa
Ultrahigh vacuum	$\leq 10^{-7}$ Pa

VACUUM PUMPS COMMONLY USED IN EM LABS

Having established briefly what a vacuum is, the next consideration is how to create the vacuums needed to operate electron microscopes and EM specimen preparation equipment. Vacuum pumps work by two broadly different functional principles. The first manner of operation is to transfer the gas physically from the volume to be evacuated to an area of higher pressure. Pumps that work by this principle are called mechanical pumps. Most of the pumps used in conventional electron microscopes and sample preparation equipment are mechanical pumps. A second type of vacuum pumping process is to sequester or entrain the gas, by some physical or chemical means, within the volume to be evacuated. The gas molecules are still present in the vacuum chamber, but cannot contribute to the gas pressure. These types of pumps are called entrainment pumps. They are found mainly on newer, higher-vacuum devices.

Rotary Pumps

Rotary-vane vacuum pumps are common to nearly all EM equipment. Although their final vacuum is not adequate for most EM laboratory applications, they do serve two important functions. First, they can be used to lower the pressure in the vacuum system (rough-pump) to the point that other types of pumps, that produce much better final vacuums, can work efficiently. Second, they provide a means of removing gas from the outlet of higher-vacuum pumps that cannot efficiently discharge the gases that they pump to atmospheric pressure. Pumps that handle this task are called backing (or fore) pumps.

A rotary-vane pump consists of a rotor positioned eccentrically in a chamber with spring-loaded vanes that contact the chamber walls (Figure 3.1A). As the rotor turns, the space between the rotor and the chamber walls (which is partitioned by the vanes) increases. At this point of the cycle the partitioned volume near the inlet port is increasing and thus has a lower pressure than that at the exhaust port, thereby drawing gas in. As the next vane passes the inlet port, the volume is closed off and compressed toward the exhaust port, where the transferred gases are expelled. Backward flow of gas is prevented by a one-way valve located at the exhaust port. The whole pump system is sealed and lubricated by an oil bath. For most EM applications a highly purified mineral oil affords the best combination of sealing, lubrication, and heat-transfer characteristics.

The rotary pump can be staged so that the first stage exhausts into the inlet of a second rotary pump. Such a setup lowers the minimum pressure that can be achieved (Figure 3.1B). Normally, the two stages are arranged coaxially, so the rotors share a common shaft.

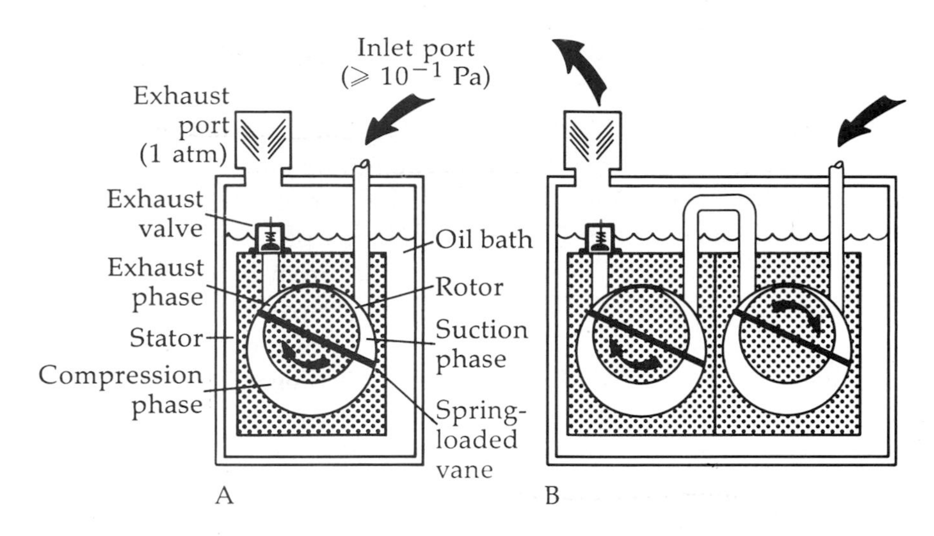

Figure 3.1 Rotary pumps. (A) Schematic of a rotary-vane pump. (B) Schematic illustrating the concept of staging. The two stages are usually mounted coaxially to decrease the number of moving parts. Rotary pumps are used more frequently than any other type of vacuum pump in the EM lab.

Pumping characteristics of rotary pumps are usually expressed in terms of their free-air displacement and their ultimate vacuum. As the vacuum improves in the volume being evacuated, the pumping speed declines rapidly near the ultimate vacuum of the pump.

For the types of applications found in EM laboratories, two-stage rotary pumps with ultimate vacuums of about 0.1 Pa and capacities of 100 L/min to 1,000 L/min (pumping speeds are given at free-air conditions) are used.

Vapor-Diffusion Pumps mech

The vapor-diffusion pump is the most widely used type of high-vacuum pump in EM applications. Although there are more modern types of high-vacuum pumps, each with some distinct advantages over the vapor-diffusion pump, its simplicity of operation and high degree of reliability probably will keep the diffusion pump in use in EM applications for some time. The name "diffusion pump" is not, however, the most informative name for this type of pump, since it works by momentum transfer rather than by diffusion, and although there are no moving parts, it is a mechanical pump.

The vapor-diffusion pump works by directing the movement of gas molecules into the pump by collision with vapor molecules traveling at supersonic speeds. The vapor molecules are generated in a boiler at the base of the pump and rise through a stack (or chimney) to the level of an annulus with several orifices that direct the molecules to the pump wall. The pump wall in a vapor-diffusion pump is a water-cooled (or occasionally air-cooled) surface that condenses the oil vapor and provides a surface on which the oil can flow back down to the boiler region for reheating.

As the supersonic oil vapor molecules leave the jets, any gas molecule they hit is directed down the pump bore. Molecules thus can diffuse into the pump but not out of it. This process accumulates gas molecules below the first annulus and its series of jets. Below the first annulus are successive stages that continue to compress the gas molecules until they reach a pressure at which a rotary pump can easily remove them. Most pumps have three or four such annuli and can accumulate a tenfold increase in pressure across each of them (Figure 3.2).

There are two critical pressures for operating diffusion pumps. First, for the vapor molecules to reach the pump walls at supersonic speed, the inlet pressure must be below some specific value, usually about 0.1 Pa. If the pressure exceeds this value, the vapor jets can no longer maintain the necessary speed. The lower jets fail at successively higher pressures. A second critical pressure is that at the pump outlet. The backing pump must maintain a pressure of 10 Pa or less to keep reverse flow from causing the jets to fail because of accumulated gas.

Exceeding the critical outlet pressure is rare except in the case of mechanical failure (e.g., rotary pump failure). Exceeding the critical

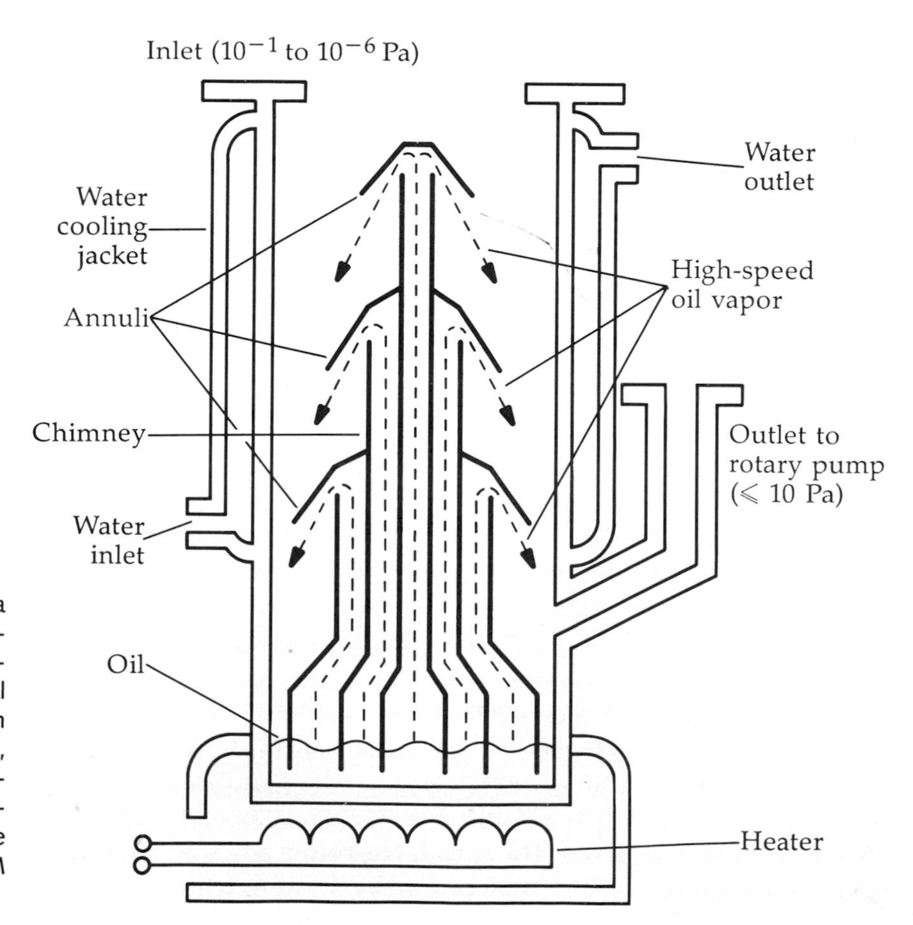

Figure 3.2 Longitudinal schematic of a three-stage vapor-diffusion pump. Although it has no moving parts, the vapor-diffusion pump is considered a mechanical pump. Gas molecules are knocked down stage by stage by supersonic oil vapor, which condenses on the cooled outer surface and flows back to the boiler for recycling. The vapor-diffusion pump is the most common high-vacuum pump in EM equipment.

inlet pressure is more common (especially with older, manually operated equipment), where rough pumping is liable be done incompletely.

The pumping speed and ultimate vacuum attainable with a vapor-diffusion pump depend on the pump size and design and the type of fluid. For a given pump design, the pumping speed increases roughly as the square of the pump diameter. Pumps for EM applications have pumping speeds in the range of 100 L/s to 1,500 L/s.

Although mercury was the first pumping fluid used in vapor-diffusion pumps, organic fluids are currently used for most applications because of their much lower vapor pressures and lower toxicity. Hydrocarbon oils with low vapor pressures give satisfactory performance in some diffusion-pumped vacuum systems. They are the least expensive but suffer from a shorter working life and higher reactivity with oxygen, which makes them rather unforgiving in the event of improper pump operation. Silicon-based fluids are considerably more forgiving to vacuum problems but can polymerize if exposed to electron or ion beams. A class of long-lived fluids with superior vacuum properties for EM applications is the polyphenyl ethers (e.g., Santovac 5).

The ultimate pressure attainable with a vacuum pump, especially when water-vapor or pump-fluid fractions may be present, can be improved by the addition of a cold baffle to trap these heavier particles. In vacuum evaporators in the EM laboratory, which are regularly vented to room air, the addition of a liquid nitrogen–cooled baffle allows about a tenfold increase in the ultimate vacuum attainable in a reasonable time. Since the baffle slows the pumping rate in dry air, most of the advantage is seen when pumping air with a high water-vapor content.

Turbomolecular Pumps

The turbomolecular pump is a second type of high-vacuum pump that transfers gas by mechanical means. As a source of high vacuums, turbomolecular pumps have several advantages over vapor-diffusion pumps. The main advantage of a turbo pump is its ability to provide a clean vacuum because of the absence of the backward diffusion of oil that occurs in vapor pumps operating near their ultimate vacuum. An additional advantage is that turbo pumps have no need to warm up, as a diffusion pump must, prior to use. Turbo pumps for the EM and related laboratory equipment generally reach operating speed in about a minute.

A turbomolecular pump functions like an axial compressor (such as one might find in a jet engine), but at the molecular-flow vacuum level (Figure 3.3). The pump consists of a stack of slotted rotors that rotate at speeds as high as 90,000 rpm. Between the rotors are stationary slotted plates (called stators). The slots in the rotors and the stators are angled in such a way that a gas molecule hit by the rotor will more likely

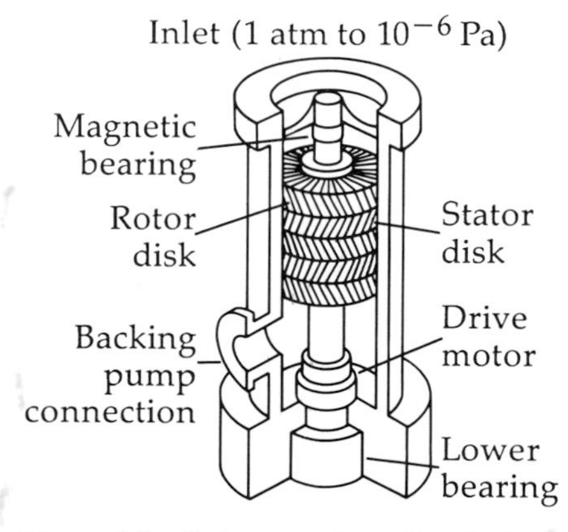

Figure 3.3 Cutaway schematic of a turbomolecular pump. The turbo pump rotor/stator sets act as compressor stages, each concentrating the gases more. These pumps require a (usually rotary) backing pump to remove gas from the last stage. Turbo pumps provide a clean vacuum rapidly.

travel to the next stage than be ejected back into the vessel being pumped. The successive rotor/stator sets act as compressor stages, each one concentrating the gas more (analogous to the successive annuli in the diffusion pump). Typical turbo pumps have five to fifteen of these stages, with the rotors mounted either horizontally or vertically.

In addition to the advantages of turbomolecular pumps, some drawbacks arise from the practical aspects of their operation. Although the turbo pump can create vacuums from air in the 10^{-9} Pa range, the efficiency for lighter gases is reduced. At ultimate vacuum the residual gas in a turbo-pumped system usually is about 99% hydrogen. Turbomolecular pumps can work effectively only when the drag on the rotor assembly is lowered, a condition accomplished by reducing the exhaust pressure with another vacuum source, usually a rotary pump.

Modern turbomolecular pumps have overcome many of the technological problems that troubled earlier pumps. The vibration of the high-speed rotors has been reduced greatly, and effective vibration dampers are available.

Getter Pumps

Getter pumps are in the class of high-vacuum pumps that entrain gas molecules within the chamber to be evacuated rather than transferring them to the outside. Titanium-sublimation pumps (TSP) are ultrahigh-vacuum pumps that work by providing a fresh, reactive surface for the gettering process. A getter surface is one on which molecules of reactive gases (such as hydrogen, oxygen, nitrogen, and water vapor) form sorptive and chemical bonds that trap the molecules to the surface. These surfaces are usually cooled by water or liquid nitrogen to improve pumping speed. After some period of time, the surface is renewed by the sublimation of a fresh coat of gettering metal, which coats over the previously trapped molecules.

The TSP is a fairly simple device consisting of a cylindrical chamber with a centrally located filament assembly. Typically the filament assembly has several sequentially used filaments that extend the service life of the unit (Figure 3.4).

The filament material is usually a titanium alloy with 15% molybdenum added to improve its sublimation rate and increase its life. During a regeneration cycle, the filament is heated to the sublimation point, and a fresh film of metal is deposited on the pump chamber walls, forming the gettering surface and burying the sorbed molecules.

Since the TSP requires that a reactive metal be heated to the sublimation point, the initial vacuum must be quite good. At pressures above about 0.1 Pa the heated filament reacts with the gas in the system and does not sublime. At vacuums in the 10^{-4} Pa range, the filament usually needs to be run continuously to keep the gettering surface fresh and the pumping speed high. In the normal operating range for the TSP

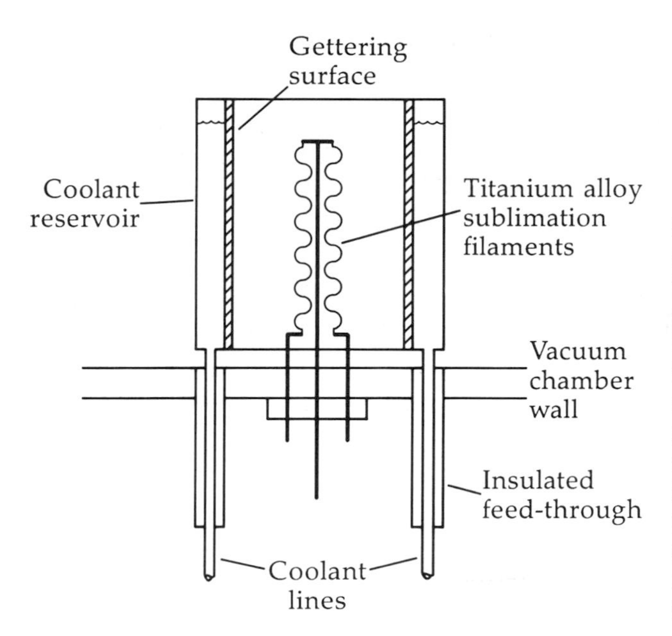

Figure 3.4 Longitudinal schematic of a titanium-sublimation pump. Titanium-sublimation pumps are found on ultrahigh-vacuum systems. Gas molecules stick to the walls of the pump body freshly coated with titanium. These surfaces are much more efficient at lower temperatures.

($< 10^{-4}$ Pa), pumping speeds are determined largely by the size of the pump body, i.e., the size of the getter surface. Typical units have capacities of 1,000 L/s to 10,000 L/s, and the rate is nearly doubled by running the pumps at liquid-nitrogen temperatures rather than at room temperature (Table 3.3).

The major application of TSPs in the EM lab is to generate ultrahigh gun-chamber vacuums for microscopes that use the vacuum-sensitive field-emission guns. In these instruments, one or more pumps are situated near the gun chamber and complete the pumping after a more traditional vacuum system has evacuated the microscope column.

Sputter-Ion Pumps

A second class of ultrahigh-vacuum entrainment pump is sputter-ion pumps. The basic operating principle of sputter-ion pumps is that ions are more reactive with the entraining surfaces than are neutral

Table 3.3
Approximate Specific Pumping Speeds for a Freshly Coated Titanium Surface at $< 10^{-5}$ Pa

	PUMPING SPEED (L/s cm ^{2})					
Titanium Surface Temperature	H_2	N_2	O_2	CO	CO_2	H_2O
20°C	3	3	9	10	5	3
−196°C	9	9	11	11	9	14

atoms; thus, an ionized gas more easily pumped. Within this class of pumps, several approaches have been used to generate the ions and extract the most pumping effort. As with sublimation pumps, the vacuums produced by ion pumps are very clean. Ion pumps are more efficient at pumping noble gases (e.g., helium) than are TSPs.

An ion pump consists of a cathode and collector system similar to a diode or triode vacuum tube. The central anode assembly consists of an array of stainless steel cylinders, the axis of which lies along an externally applied magnetic field (Figure 3.5). This magnetic field keeps the emitted electrons from quickly reaching the anode surface, and they are thus trapped in the stainless steel cylinders at high concentrations. Incoming gas molecules are ionized. Since their mass is much greater than that of the electrons, they are largely unaffected by the magnetic field and hit the cathode of the cold-cathode discharge system. Upon impact, the ions sputter out some of the titanium cathode material, which forms a getter surface on the pump walls that is effective in trapping reactive gases. In addition to the gettering effect, some of the ions are buried directly by their impact momentum, some molecules are buried by the sputtered cathode material, and some neutral molecules formed by dissociation or neutralization of ions are bound to the walls by adsorption or burial. The manner of gas entrainment depends on the type of gas being pumped and how the gas behaves in the ionizing environment.

Sputter-ion pumps are used in ultrahigh-vacuum applications in which a clean (oil-free) vacuum is needed. The initial vacuum may be

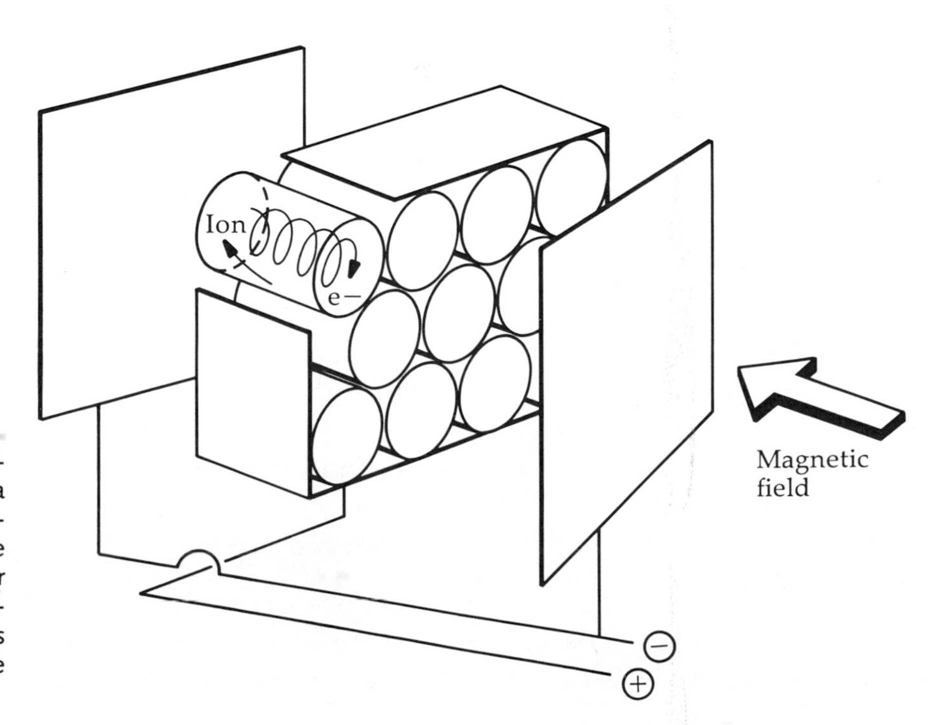

Figure 3.5 Schematic of a cell of a sputter-ion pump. The ion pump consists of a cathode and a collector system. The magnetic field greatly increases the flight time for the ionizing electrons. Because of their much greater masses, ions are not deflected into the tight coils that electrons are, but fly directly to the collector. The ions lose their charge and are entrained.

provided by diffusion or turbomolecular pumps. Often they are used in conjunction with sublimation pumps to form a more efficient system. Sputter-ion pumps are essentially maintenance- free for the life of the cathode, which is eventually sputtered away after several thousand hours (depending on the range of vacuums pumped). Since there is a fairly strong magnetic field around the pump, care must be taken in its orientation and shielding around magnetically sensitive instruments, such as electron lenses. Sputter-ion pumps are used as gun-chamber pumps on newer field-emission SEMs and many TEMs.

Cryoadsorption Pumps

Cryoadsorption pumps are a third class of entrainment pump that works on sorptive principles. The cryoadsorption pump works by condensing some of the gas molecules out of the vacuum chamber onto a cold surface and trapping others in cold, adsorptive material. As with titanium-sublimation pumps and ion pumps, cryoadsorption pumps have no oil-containing systems exposed to the vacuum and, thus, produce very clean vacuums. In addition, the absence of moving parts in the pump body makes cryopumps vibration-free.

Although there are several approaches to construction of a cryopump (the "cold fingers" used in TEM anticontamination devices are perhaps the simplest), the more common type for use as a vacuum pump is the two-stage helium-gas refrigerator-cooled cryoadsorption pump. The cryoadsorption pump consists of a chamber or pump body within which are two adsorptive stages (Figure 3.6). The first stage consists of a radiation shield kept at about –190°C that traps water vapor and protects the second stage from gross contamination and radiant heating. This stage may be cooled by liquid nitrogen to reduce the load on the refrigerator system.

Within the radiation shield is a second stage consisting of a cold surface with an adsorptive material, such as activated charcoal, fixed to one surface. This stage is kept at about –250°C, at which temperature all gases except neon, hydrogen, and helium are frozen. The neon, hydrogen, and to a lesser extent helium are adsorbed by the adsorptive material fixed to one side of the second-stage head.

The cryopump requires a heat sink to remove heat added to the adsorptive surfaces by molecular collision and condensation and by radiant heating. As mentioned earlier, the first stage can be cooled by liquid nitrogen, which can be vented to air. The second stage, requiring a temperature of about –250°C, can be similarly cooled by liquid helium. Since helium is a relatively expensive and finite resource and is more difficult to handle in its liquid form, closed-system refrigerators that recycle the helium are now used. These units consist of a compressor that compresses the helium to a few megapascals, a heat exchanger that brings the gas back to room temperature, and a gas

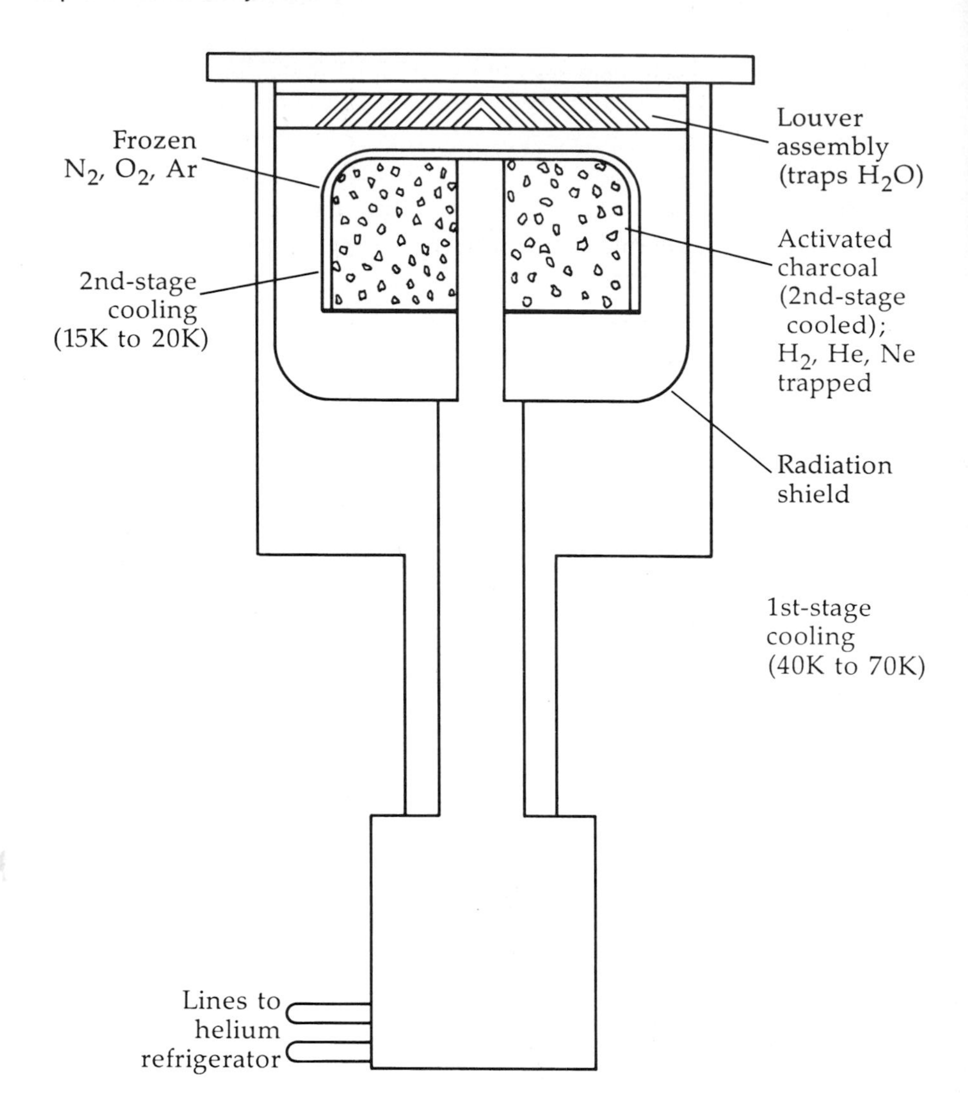

Figure 3.6 Longitudinal schematic of a cryoadsorption pump head. The two-stage cryoadsorption pump uses two cooling ranges either to freeze or to adsorb gas. The first stage traps water vapor and shields the inner stages. The second stage traps all gases either by freezing them or by adsorbing them.

cleaner. This cool, high-pressure gas is returned to a two-stage expansion head where the pressure is reduced, cooling the pump head in the process.

Since cryopumps have a finite capacity, their operation requires two considerations. First, since the gas that they pump is trapped on a surface, they function only as long as the accumulation of material on the adsorptive surfaces is small. The amount of accumulation is reduced by rough-pumping the system with a rotary pump. Second, when the adsorptive surfaces become loaded with condensed gas, their thermal conductance and pumping efficiency are reduced. They must, therefore, be brought up to room temperature periodically to allow the trapped gas to evaporate and be rough-pumped away. Under normal conditions, this event occurs about once a week. Cryopumps have high pumping rates for most gases of concern in EM applications. They can achieve a pressure lower than 10^{-10} Pa in air and are especially effective at pumping water vapor. How effectively they can pump a specific gas is related to the temperature attained at the cold head.

METHODS OF MEASURING VACUUMS

To use vacuum pumps effectively in the EM laboratory, gauging systems are needed to measure the level of the vacuum. Gauges also control valves and pumps in automated systems. The high and ultra-high vacuums required in EM applications are beyond the range and accuracy of direct-reading mechanical gauges, such as mercury manometers and diaphragm types. The types of gauges used in most EM lab applications are indirect-reading gauges. These instruments calculate the gas pressure by measuring some pressure-related property of the gas and transducing that property to a measurable electronic signal. Four types of vacuum gauges are commonly used in EM equipment. These gauges and their effective ranges are shown in Table 3.4.

Pirani and Thermocouple Gauges Measure Low Vacuums

The Pirani gauge is an indirect-reading gauge that is useful for measuring the vacuum in a system between about 100 Pa and 10^{-1} Pa. Pirani gauges work on the principle of heat loss from a hot wire to the surrounding residual gas (Figure 3.7).

The gauge system consists of two resistor wires, one maintained under standard conditions and the other exposed to the vacuum environment. A current run through the two resistor wires heats them. As each wire heats up, some of the heat is dissipated by gas molecules colliding with the wire. The more molecules that hit the wire, the greater the heat loss from the wire and the lower its temperature. As the temperature decreases, the resistance value drops, and the current through the wires increases. Thus, the reference wire, held at constant pressure, has a fixed resistance, whereas the resistance of the wire being

TABLE 3.4
Vacuum Gauge Ranges

Type of Gauge	PRESSURE RANGE (Pa)							
	10^4	10^2	10	10^{-2}	10^{-4}	10^{-6}	10^{-8}	10^{-10}
Thermocouple	-----	-----	-----	-----	-----			
Pirani	-----	-----	-----	-----	-----			
Cold-cathode				-----	-----	-----	-----	
Hot-cathode					-----	-----	-----	-----

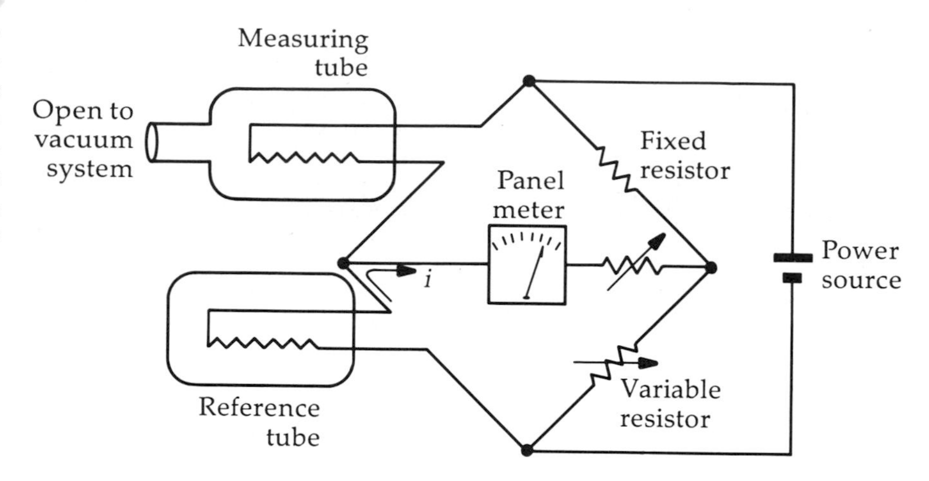

Figure 3.7 Pirani gauge system. Schematic of the gauge tubes and the measuring circuit. As heat is carried away from the measuring tube filament by gas molecules, the Wheatstone-bridge circuit registers a current. The more gas molecule collisions, the cooler the wire and the greater the current, which is read directly as a vacuum on the instrument panel.

measured decreases as the vacuum increases. The difference in current between the measuring-wire side and the standard side of the bridge is detected by an ammeter connected in a Wheatstone-bridge circuit, and the vacuum is determined from this reading of the current.

The thermocouple gauge is another type of indirect-reading vacuum gauge. As with the Pirani gauge, the thermocouple gauge uses heat flow from a wire to measure the vacuum. Instead of comparing the vacuum-environment resistor with a standard, however, the thermocouple gauge has a small thermocouple (e.g., of copper-constantan) welded to the middle of the wire. As the vacuum and the temperature of the wire increase, the change in current is read on a millivolt meter attached to the thermocouple lead (Figure 3.8).

Both the thermocouple and the Pirani gauges are reliable. The thermocouple is perhaps more rugged but is generally less accurate. These gauges are used on EMs and related specimen preparation equipment to monitor low to medium vacuums. A typical EM system uses these gauges to measure the vacuum from air to until rough-pumping is completed. When the system is switched to the final high-vacuum pump, the Pirani or thermocouple gauge is turned off, and a high-vacuum gauge is started.

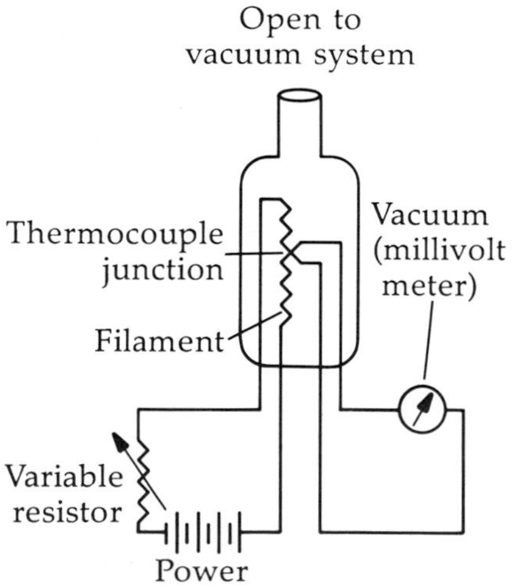

Open to vacuum system
Thermocouple junction
Vacuum (millivolt meter)
Filament
Variable resistor
Power

Figure 3.8 Thermocouple gauge system. Schematic of the gauge tube and the measuring circuit. The thermocouple gauge simply measures the temperature on a heated wire with a thermocouple, which is attached to it. As the vacuum increases, the wire gets hotter because fewer gas molecules are colliding with it and removing heat.

Cold-Cathode (Penning) and Hot-Cathode Gauges Measure High Vacuums

The type of high-vacuum gauge most commonly used in EM applications is the cold-cathode, or Penning, gauge. When the gas pressure in the volume being evacuated becomes so low that the gas flow is in the molecular range, the thermal conductivity effect becomes too slight to use it for gauging the vacuum. The cold-cathode gauge amplifies the electronic effect of a gas molecule by ionizing it and increases the chance of ionizing other molecules by lengthening their paths with magnetic fields (Figure 3.9).

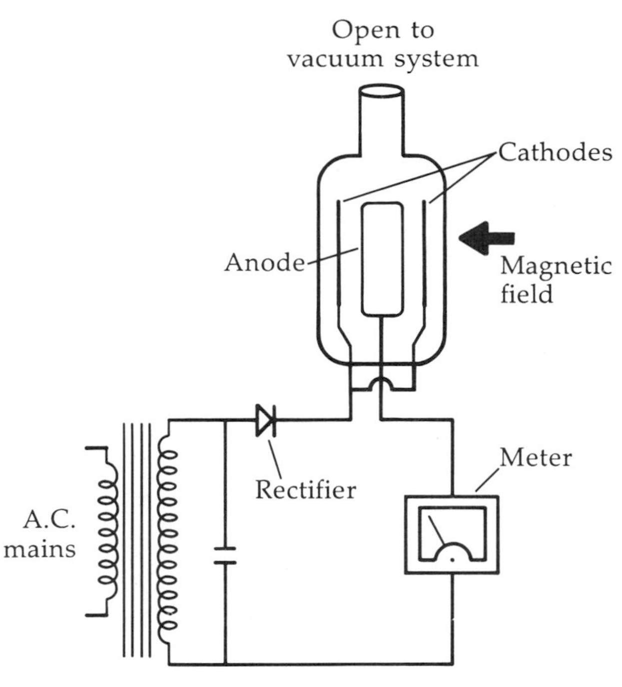

Figure 3.9 Cold-cathode gauge system. Schematic of the gauge tube and the measuring circuit. The cold-cathode gauge works by a magnetic field that lengthens the time of flight of electrons emitted from a cold source, improving the chance that they will hit a gas molecule and ionize it. Gas molecules ionized by these collisions create a current proportional to the vacuum.

The general layout of the cold-cathode gauge is similar to that of the ion pump mentioned earlier (see Figure 3.5). Cold-cathode gauges have considerable pumping potential because of the same kind of sputtering that occurs in the ion pump. Thus, the openings into the gauge tube are made as wide as possible, and vacuum lines to the gauge should be as wide and straight as possible to avoid having the gauge read a vacuum that it itself has made artifactually high.

The Bayard-Alpert hot-cathode vacuum gauge is another type of ionization gauge that extends the measurable vacuum range about two to three orders of magnitude below what can be measured reliably with the Penning gauge. In a hot-cathode gauge, the cathode wire is heated to the thermionic emission point, at which a driving potential of about 200 V accelerates the electrons toward the anode. Gas molecules entering the body of the hot-cathode gauge are ionized by these electrons and preferentially driven to an ion collector that is held at a more negative potential than that of the cathode. Upon collision with the collector, the ions lose their charge, causing a current to be generated, which is amplified and displayed as the corresponding voltage on a meter calibrated in vacuum units (Figure 3.10).

With modern gauge-head design and electronics, these gauges can monitor vacuums into the 10^{-10} Pa range. Hot-cathode gauge systems are more expensive than cold-cathode gauge systems, and the incandescent gauge heads are much more labile to damage from vacuum faults.

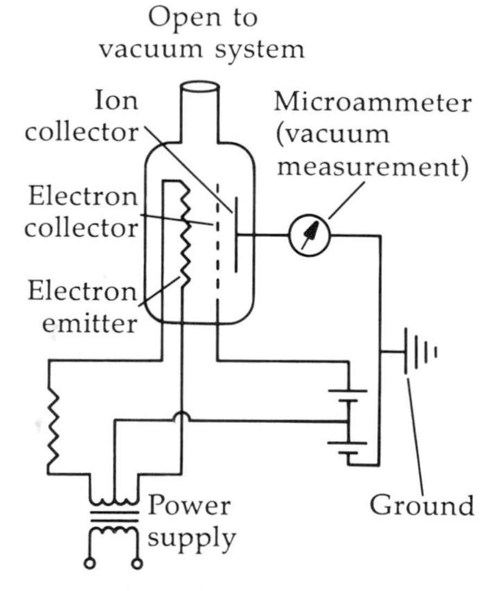

Figure 3.10 Hot-cathode gauge system. Schematic of the gauge tube and the measuring circuit. At higher vacuums, the cold discharge of electrons is insufficient. By heating the electron emitter, the electron flux improves, providing a higher probability that gas molecules will be ionized and eliminating the need for a magnetic field. Hot-cathode gauges can measure higher vacuums but are more sensitive to leaks than cold-cathode gauges are.

VACUUM SYSTEMS USED IN ELECTRON MICROSCOPY

Two-Stage Rotary/Oil Diffusion–Pumped Systems

The two-stage rotary / oil diffusion–pumped high-vacuum system (Figure 3.11) is at present the most common. Most electron microscopes, both scanning and transmission, are equipped with this kind of system.

The first stage consists of a rotary rough-pump, which is used to pump the microscope column down to the vacuum at which the diffusion pump can operate. When this vacuum is reached, as determined usually by a thermocouple or Pirani circuit, the valve between the rotary pump and the system is closed, and the valve isolating the diffusion pump from the system is then opened. The evacuation progress of the diffusion pump is monitored with a Penning gauge. Generally, the gauge signals in modern EMs are monitored electronically and used by various logic circuits to control the opening and closing of valves in an automated system.

To eliminate some of the vibration from the rotary pump, some manufacturers incorporate a backing tank in the system. This vessel serves as a reservoir of vacuum to back the diffusion pump and allows the rotary pump to be switched off, which is especially useful during micrography. When the gauge monitoring the vacuum in the tank reads high again, the rotary pump is switched on to reevacuate the tank. Alternatively, some manufacturers mount the rotary pump separately from the microscope and use vibration dampers. Similar vacuum systems are used (especially in older models) in EM specimen preparation

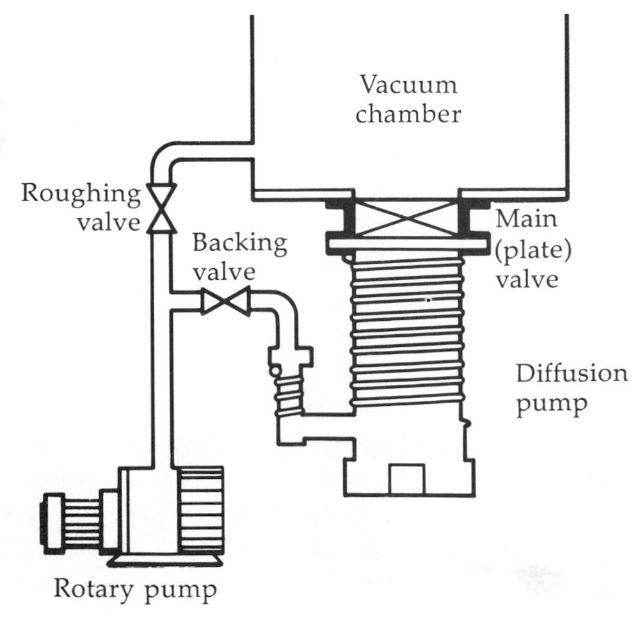

Figure 3.11 Diffusion-pumped vacuum system. This vacuum system is the one found on most electron microscopes. The rotary pump first rough-pumps the chamber or column; then the final vacuum is attained by opening the main valve to the diffusion pump. The rotary pump then backs the diffusion pump by removing gas compressed in the space of its bottom stage.

devices such as vacuum evaporators, ion mills, and freeze-etch equipment. When a diffusion-pumped system is used in an ultrahigh-vacuum operation, the system may have provisions for either isolating the diffusion pump and switching to an ion or sublimation pumping system or differentially pumping the area of the system requiring the ultrahigh vacuum.

Diffusion-pumped systems have several advantages. First, with periodic preventive maintenance, a diffusion-pumped system will last the life of the equipment it is used in. Second, for their size, diffusion-pumped systems can pump a large quantity of gas rapidly. Third, even with considerable contamination they still function.

Diffusion-pumped systems also have a few disadvantages. A major problem in many EM applications is the back-streaming of oil vapor. Although the diffusion-pump fluids have very low vapor pressures, they nevertheless can make up a considerable portion of the residual gas in a diffusion-pumped system. In some applications, such as freeze etch, this can pose serious contamination problems. Although this problem can be nearly completely alleviated by the addition of a baffle on the pump, the baffle also slows the pumping speed. A second problem is that a diffusion-pumped system requires a constantly running rotary pump or vacuum buffer, both for the rough-pumping of the system and to remove the gas from the diffusion-pump body. Since the fluid return in a diffusion pump is operated by gravity, the pump needs to be mounted in a stable vertical position. Whether the system is switched from rough-pumping to diffusion pumping manually or automatically, the system involves at least three valves, which must be operated correctly, in the pumping sequence. Finally, the diffusion pump needs to be warmed up before and cooled down after use for about a half hour.

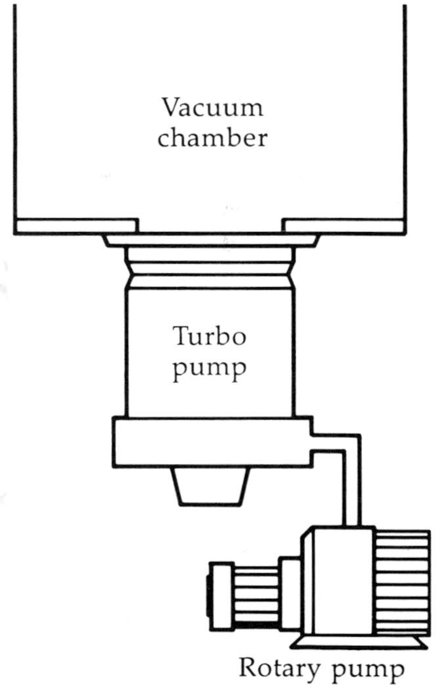

Figure 3.12 Turbomolecular-pumped vacuum system. The turbomolecular-pumped system can be valveless, since the rotary backing pump can rough-pump the system through the accelerating turbomolecular pump. The turbo pump needs to be backed by the rotary pump to lower drag on the rotors to the point at which the turbomolecular effect can take place.

Turbomolecular-pumped Systems

The turbomolecular-pumped system is conceptually the simplest system and is now available on some SEMs (standard equipment on at least one manufacturer's line) and on a wide assortment of evaporators and other specimen preparation equipment. Turbomolecular pumps can begin operation from ambient pressure and thus require no system prepumping. The turbomolecular pump gains speed as the bulk of the gas is removed, a process that is generally completed quite quickly. All that is required is the turbo pump with the rotary backing pump attached to its exhaust (Figure 3.12).

Turbo-pumped systems have a number of advantages and disadvantages compared to the diffusion-pumped system. They produce a clean vacuum. If run properly, no significant quantity of lubricant or pump oil is exposed to the vacuum to contaminate specimens or electron-optical components. They also are fast; turbo pumps require no preheating or rough-pumping operations, since the backing pump essentially rough-pumps the system, while the turbo pump is coming up to speed. Finally, the turbo-pumped systems do not need any of the valves or operating electronics required for the diffusion-pumped systems. They also have a low operating cost for routine use.

Turbomolecular pumping systems have some drawbacks that make them less suitable for some applications. Their initial cost is higher than that of diffusion-pumped systems. As with the diffusion systems, the turbo-pump needs a full-time rotary backing pump. If foreign material is drawn into the rotor chamber while at maximum speed, catastrophic damage to the pump may result. Some pumps require periodic lubrication of the rotor-shaft bearing to prevent bearing failure and destruction of the pump. Finally, if the pump system is not mounted optimally, vibrational or electrical interference may result.

Closed-System Cryopumped Systems

The closed-system cryopumped system is a more recent development in EM applications than the other vacuum systems are. The basic layout of the system is similar to that of the diffusion-pumped system, with the exception that once the system has been rough-pumped by the rotary pump, it can be shut off for the rest of the vacuum cycle (Figure 3.13).

In general, these systems are used with customized analytical electron microscopes, where the possible contamination from a diffusion pump and the environmental noise of a turbomolecular pump cannot be tolerated. They are now also available as the vacuum source in the best freeze-etch machines.

Since cryopumped systems have no pumping oils or mechanical parts exposed to the vacuum, they are very clean. They can be mounted in any orientation in the system and have high speeds, especially for water vapor. They do, however, have a high initial cost and require periodic regeneration (outgassing of trapped gas) and maintenance (addition of high-purity helium, etc.). If the helium supply to the pumps is interrupted by power failure, the system will probably need to be regenerated completely.

Most vacuum systems in use for EM applications are one of the three types already described: rotary/oil diffusion–pumped, turbomolecular-pumped, or cryopumped. They may work individually or together with other equipment. To achieve optimum performance it is wise to consult the operating protocols provided by the manufacturer.

Differentially Pumped Systems

For some EM applications, it may not be necessary or even desirable to have a uniform vacuum in the system. In the simplest case,

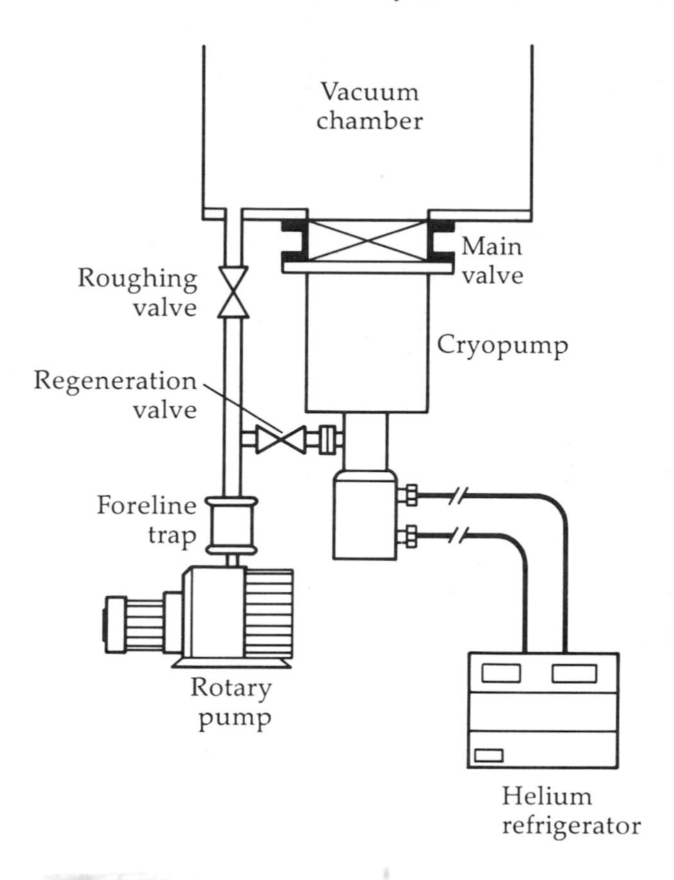

Figure 3.13 Cryo-pumped vacuum system. The cryopumped system requires a roughing pump, much like the diffusion-pumped system. Once the bulk of the gas has been rough-pumped out of the system and the plate valve to the cryoadsorption pump opened, the backing pump is no longer needed, since the gas is entrained in the cryopump head. As the adsorptive surfaces become saturated, the pump needs to be regenerated by heating the pump head and pumping out the liberated gas with the rotary pump.

preparation equipment such as sputter coaters and glow-discharge units have controlled leaks that allow the pumping system to create a partial vacuum of known composition (e.g., argon). In microscopes with field-emission guns, the gun chamber is kept at an ultrahigh vacuum by sputter-ion and/or sublimation pumps, while the rest of the system may be at a vacuum several orders of magnitude lower. This differential vacuum is accomplished by the small diameter of the EM column, which is a major restriction to the migration of gas molecules in the molecular-flow vacuum range. A final method of differential pumping is used in the environmental SEM. In this instrument, the specimen chamber is kept at a low vacuum and the residual gas kept away from the electron-optical regions of the column by a series of small apertures that divide the lower column into a series of regions that can be pumped differentially. The space above the apertures can be kept at a vacuum thousands of times better than that of the specimen chamber.

FURTHER READING

Leybold, Inc. 1987. *Vacuum Technology: Its Foundations, Formulae and Tables.* Leybold, Inc., Export, PA.

O'Hanolon, J. F. 1980. *A User's Guide to Vacuum Technology.* Wiley, New York.

4

The Transmission Electron Microscope

The transmission electron microscope (TEM) was the first electron microscope. The TEM produces a transmitted electron image of a thin specimen, magnified from 100 to approximately 500,000 times, and with a resolving power of approximately 0.2 nm. To produce the standard bright-field TEM image (Figure 4.1), the electron beam must be able to penetrate the sample, with many electrons being transmitted through it. The quality of the image in the TEM depends not only on the expertise of the microscopist, but also on the quality of the specimen preparation. General procedures for specimen preparation for the TEM are described in Chapter 6.

In biology a TEM is used to obtain high-resolution images of internal cellular structures of animal and plant tissues, as well as microorganisms, and of the structures of small particulate specimens such as viruses, phages, and DNA. With special specimen preparation procedures, the TEM can also be used for the localization of elements, enzymes, and proteins; the study of membrane interfaces; and the study of the structure of macromolecules.

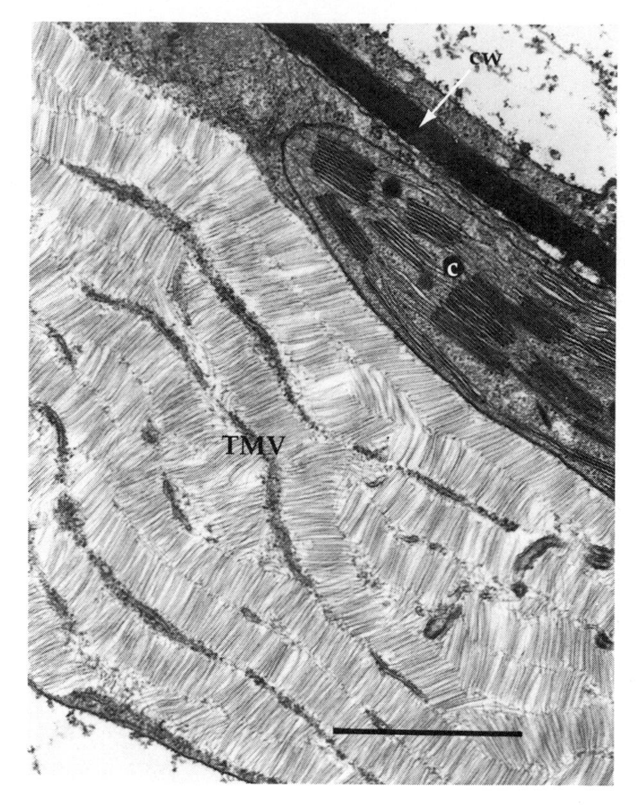

Figure 4.1 Transmission electron micrograph. Ultrathin section through a tobacco leaf infected by tobacco mosaic virus (TMV). Leaf tissue was fixed in glutaraldehyde and osmium tetroxide and embedded in ERL epoxy resin. Chloroplast (c) and plant cell wall (cw) are also visible. Bar = 1 μm.

In materials sciences, the TEM is used to examine interfaces in composite materials, dislocations in metals, the fine structure of polymers, thin metal films, the crystal lattice structure of catalysts, layered clays, and other specimens and to localize elements within specimens.

THEORY OF OPERATION

The fundamental principles of optics apply to the TEM imaging process. As described briefly in Chapter 1, the resolving power of a TEM is related to the wavelength of the energy source used to form the images. De Broglie defined the wave nature of electrons and the variables that affect wavelength. His equation states: $\lambda = h/mv$, where λ is the wavelength, h is Planck's constant, m is the mass of the particle, and v is the velocity of the particle. From this equation it is easy to see that the wavelength of a particle or beam of particles can be decreased by increasing the velocity of the particles.

Abbe's equation states: resolving power = $0.61\lambda/\text{N.A.}$, where numerical aperture (N.A.) is defined as $n\sin\alpha$, where α, the aperture angle, is one-half the acceptance angle of the lens (approximately 0.3° for a TEM), and n is the refractive index of the medium (approximately 1 for the vacuum in a TEM). Numerical aperture defines the maximum cone of light that a lens can take up from a point on the specimen. Angular aperture is the angle within the lens in which the most divergent rays can pass and form an image. In general, a large angular aperture helps to reduce diffraction and therefore to improve resolution. This benefit, however, has practical limits because of the increased spherical aberration that occurs with large angular apertures.

Relating Abbe's and De Broglie's equations demonstrates that aside from lens aberrations, resolving power is limited by the wavelength of the energy source used to construct magnified images and that wavelength is related to both the velocity and the mass of the energy source. The wavelength of light is approximately 0.5 μm to 1 μm (500 nm to 1,000 nm), and the wavelength of an electron at an accelerating voltage of 50 kV is approximately 0.005 nm, a 100,000-fold difference. Increasing the velocity of electrons results in a shorter wavelength and, consequently, increased resolving power (Table 4.1). In actuality, because of constraints of lens aberrations, especially spherical aberration, the electron microscope provides a 1,000-fold improvement in resolving power over the light microscope, not the theoretical 100,000-fold difference.

Specifically, the spherical aberration varies with the cube of the numerical aperture ($d_S = 1/2C_S\alpha^3$) and the chromatic aberration increases linearly with numerical aperture [$d_C = C_C\alpha(\Delta E/E)$], where α is the aperture angle in radians, d is defined as the disk of least confusion (i.e., the area along the lens axis that corresponds to the real focal point of the lens), C is the coefficient of each aberration, and ΔE is the mean energy loss of the electrons (see Chapter 2, page 21). Both spherical and chromatic aberration thus contribute to the practical limitation of resolution in a TEM. The extent to which chromatic aberration limits resolution is strongly dependent on specimen thickness, which in turn

TABLE 4.1

Comparison of Accelerating Voltage, Wavelength, and Resolving Power for a Transmission Electron Microscope

As accelerating voltage increases, wavelength decreases and resolution decreases (improves).

ACCELERATING VOLTAGE (V)	WAVELENGTH (nm)	RESOLUTION (nm)
20,000	0.0087	0.44
40,000	0.0061	0.31
60,000	0.0050	0.25
80,000	0.0043	0.21
100,000	0.0039	0.19
1,000,000	0.00087	0.10

depends directly on the quality of the specimen preparation procedure. In a thick specimen, multiple inelastic collisions result in electrons of many different energy-loss levels that contribute to the loss of resolution. The ultimate resolving power depends not only on the instrument itself, but on the specimen preparation, especially the effect of specimen thickness on chromatic aberration. For amorphous samples, for example, the rule of thumb (devised by V. E. Cosslett) states that resolution can be no more than one-tenth the specimen thickness.

Lenses magnify specimen images by bending waves as they pass through the lens; the bent rays meet on the back side of the lens or are used as the object for another lens to magnify. By convention, diagrams of light microscope lenses may be used to explain the process of image formation by electromagnetic lenses, although in practice the two kinds of lenses operate quite differently (see Chapter 2). Two kinds of images are used in a TEM, real and virtual. Each type may be used in the construction of the final magnified image.

REAL IMAGES

Real images have two basic characteristics: They are magnified, and they are inverted (Figure 4.2). The focal point (F_2) in the back focal plane of the lens can be described as the point on the lens axis where rays parallel to the lens axis converge. The distance from the focal point to the center of the lens is the focal length of the lens. The shorter the

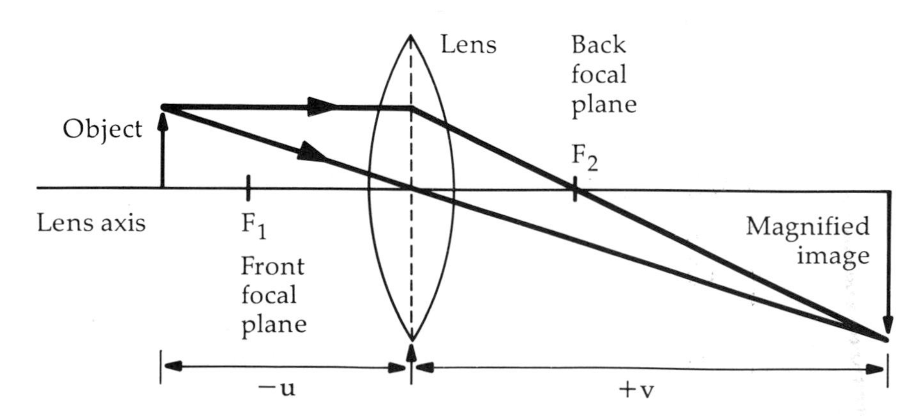

Figure 4.2 Formation of a real image. A real, magnified image is formed when the object (sample) in front of the lens is between the front focal point (F_1) and twice the focal length ($2 \times F_2$) of the lens. As the object approaches F_1, the magnification is increased. F_2 is the back focal plane of the lens. Redrawn from G. A. Meek, *Practical Electron Microscopy for Biologists*, by permission of John Wiley and Sons, New York.

focal length, the more powerful the lens. The front focal plane contains focal point F_1. The specimen must be placed in the front object plane between F_1 and $2 \times F_1$ to produce a real image in the back image plane of the lens. The distance from the specimen (object) to the lens is defined as u, while the distance from the lens to the image plane is defined as v. Magnification is calculated by v/u. Therefore, the closer the specimen is to F_1, the greater the magnification. High-magnification images are formed using real images.

VIRTUAL IMAGES

Occasionally the specimen or a specimen image (the object) is placed between the lens and the front focal point, instead of between F_1 and $2 \times F_1$ (Figure 4.3). Instead of converging to make a real image in the back image plane, the rays diverge. In this case, a virtual image is formed in the front part of the lens. It is magnified, but not inverted. The rays can be made to converge to form an image in the front focal plane if an optical system (i.e., an eye or another lens) is placed at the point of convergence. In a light microscope the eye acts as a lens to visualize a virtual image.

In a TEM, a virtual image is used to form low-magnification images. For this application, the objective and intermediate lenses are used as weak lenses. A reduced real image is formed beyond the intermediate lens, in the object plane of the projector lens. This image is a virtual image. The projector lens then forms a real image, but at a magnification that is reduced from the normal high-magnification mode.

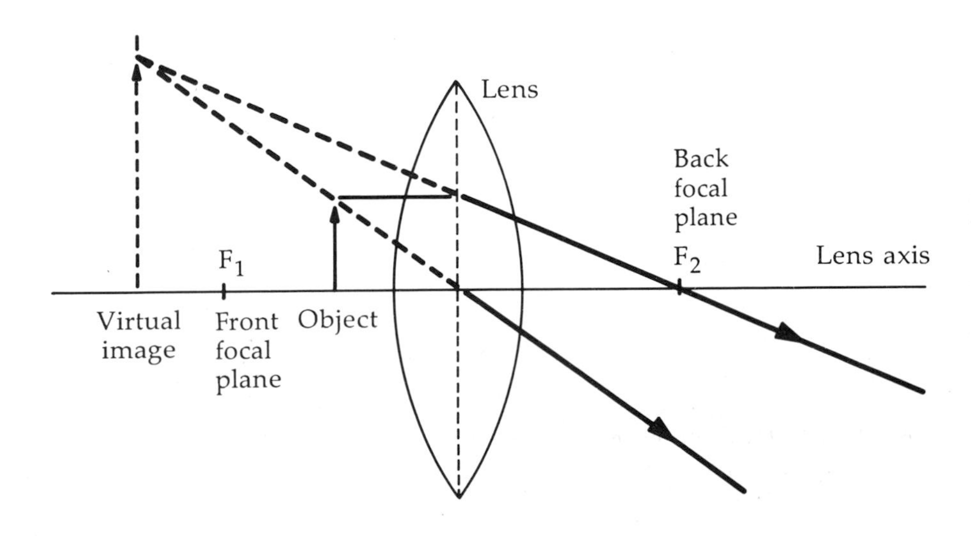

Figure 4.3 Formation of a virtual image. If the object lies between the front focal point (F_1) and the lens, the rays will diverge. In this case, no real image can be formed. Rays can be made to converge behind the object to form a virtual image, which another lens can use as an object or which an eye can visualize. Redrawn from G. A. Meek, *Practical Electron Microscopy for Biologists*, by permission of John Wiley and Sons, New York.

DEPTH OF FIELD AND DEPTH OF FOCUS

Two additional features of lenses that are important in the design of the TEM are depth of field and depth of focus. Depth of field refers to the distance along the lens axis in the object plane in which an image can be focused without a loss of clarity (Figure 4.4A). Depth of field is related to wavelength (λ) and numerical aperture (N.A.) by the following equation: $D_{field} = \lambda/(N.A.)^2$. For a typical TEM objective lens with a numerical aperture (N.A.) of 10^{-3} radians and using an accelerating voltage of 100,000 V with a wavelength of 0.0039 nm, the depth of field would be 3900 nm. Since a typical TEM specimen is less than 100 nm thick, all of the specimen is focused within the image; therefore, depth of field is not a limiting factor in TEM imaging.

Depth of focus is analogous to depth of field but refers to the distance along the axis of the image plane in which the image can be focused without a loss of clarity (Figure 4.4B). Depth of focus is related directly to magnification (M) and resolving power (R.P.) and inversely to numerical aperture (N.A.) by the following equation: $D_{focus} = M^2 \times R.P./N.A.$ For a typical TEM, in which M = 100,000, R.P. = 0.2 nm, and N.A. = 10^{-3} radians, the resulting depth of focus is 2,000 m. The depth of focus is also not a limiting factor for a TEM lens system. In fact, the great depth of focus allows one or more cameras to be placed either above or below the viewing screen with no loss of focal sharpness in any plane.

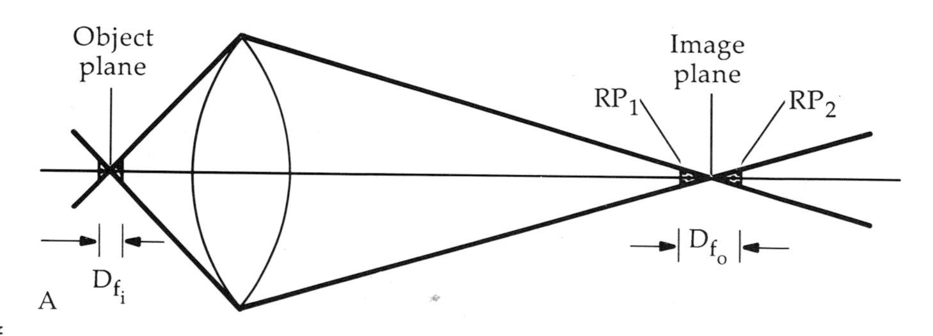

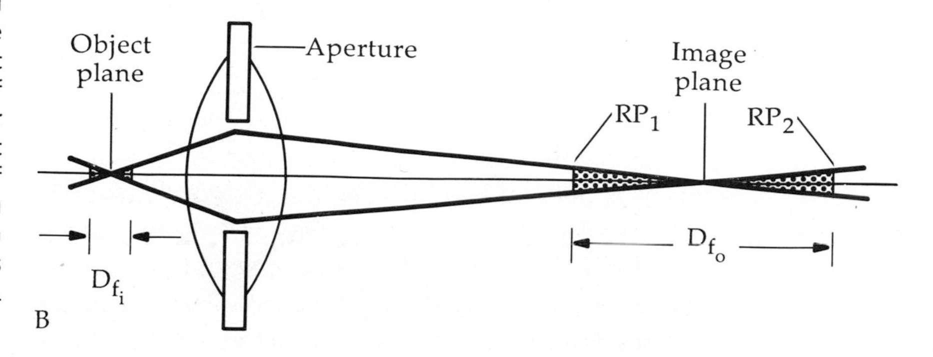

Figure 4.4 Depth of field and depth of focus of a lens. (A) The depth of field of the lens is related to wavelength and numerical aperture. It is the distance along the lens axis in the object plane that an image can be focused without loss of clarity. In a TEM this distance is larger than the standard sample thickness, so it is not a limiting factor. (B) The depth of focus is the distance along the lens axis in the image plane in which the image can be focused without loss of clarity. This distance is also not a limiting factor in a TEM.

Diffraction Is Limiting in Light Microscopy but Can Be Useful in Transmission Electron Microscopy

Diffraction can be defined as the spreading of waves into the area beyond an obstruction. As a secondary wave front is formed when a primary wave front interacts with an obstacle, it creates interference because of its differing path length (Figure 4.5). This interference then results in a blurry image, thereby decreasing resolution. The amount of diffraction depends on the wavelength of the energy source. The longer the wavelength, the greater the spreading of the secondary wave. Significant diffraction into the region behind the obstacle occurs only if the size of the obstacle is smaller than the wavelength of the energy source. For these reasons, diffraction is a major limiting factor for resolution in the light microscope.

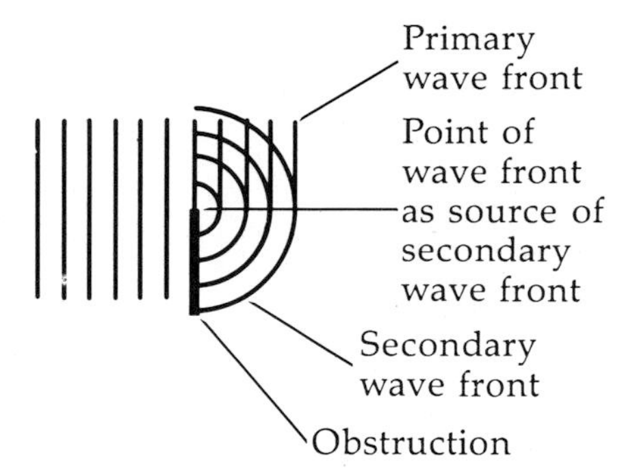

Figure 4.5 Diffraction. Any point on a wave front, where it hits an obstruction, can be the source of a new wave front. As the primary and secondary wave fronts continue to travel forward (to the right in this diagram), they interfere with each other, thereby causing a decrease in resolution. This phenomenon is very important in limiting the resolution in the light microscope but affects the TEM much less.

The British astronomer Sir George Airy demonstrated the diffraction phenomenon in the mid-1800s. Light from a point source cannot be focused through a lens to a point, but only to a disk. The "Airy disk" is a pattern of the focused light that consists of a central spot of light surrounded by several rings. The diffraction along the edge of the lens or aperture is responsible for the formation of the Airy disk. Even a perfect lens aperture or edge will cause diffraction.

Diffraction can be useful in the TEM. Fresnel fringes are one example. Fresnel (pronounced without the "s") fringes are formed when an electron beam interacts with an opaque edge. As the electron beam strikes an edge, diffraction effects result from the interaction between the unscattered beam (primary wave front) and the beam scattered by the edge (secondary wave front). This interaction results in a series of bright lines, the brightest of which is apparent at the edge of the hole, inside the hole, or outside the hole. The practical uses of Fresnel fringes are discussed and illustrated later in this chapter.

Diffraction is also useful because it contributes to contrast in images of crystalline specimens and is used to determine crystal lattice structures. These are discussed later in this chapter.

ANATOMY OF A TRANSMISSION ELECTRON MICROSCOPE

To make optimum use of the TEM, it is important to understand how the individual components function together to form a usable image. The microscopist can control a number of instrument variables, and the choices involved in optimizing the set of operating conditions affect the quality of the final image. All general descriptions in this section refer to Figure 4.6. Detailed diagrams of specific components of the TEM are included within the text that describes them. All portions

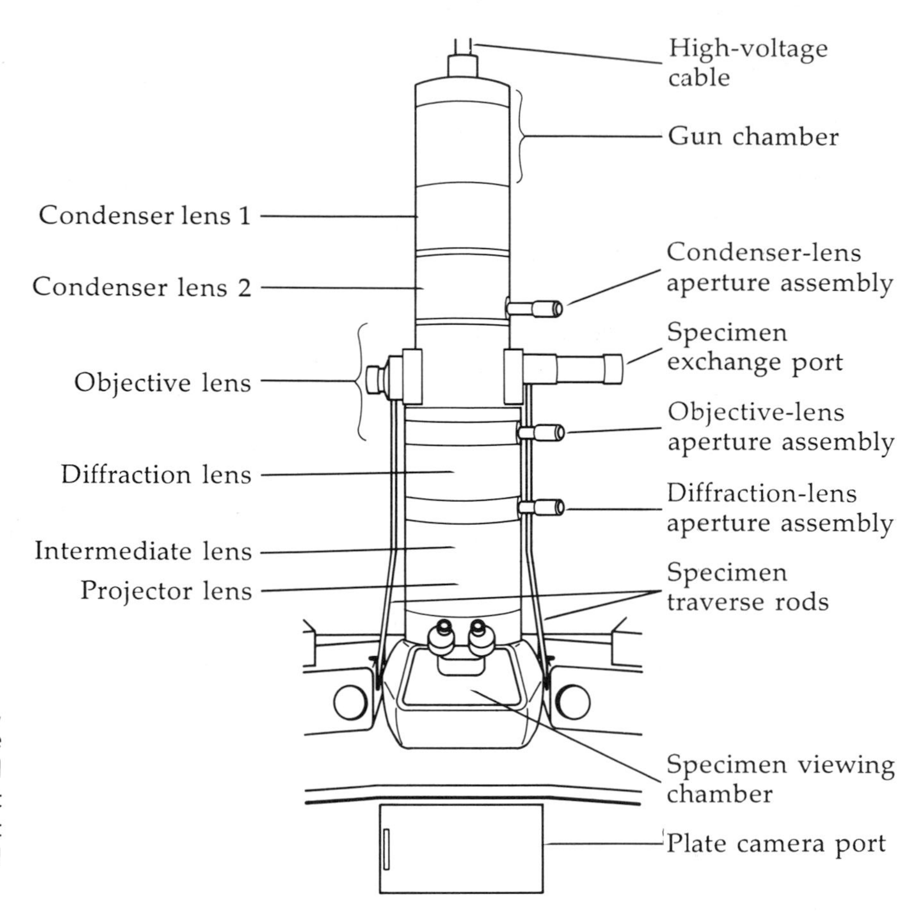

Figure 4.6 TEM column. In this view, the lenses are identified on the left of the column, and other features are identified on the right. Components of the TEM not shown are electronics, lens and filament controls, and vacuum-system lines and pumps.

of the column, from the gun chamber to the camera chamber, are under vacuum (see Chapter 3). The specific design of the lenses and the filament assembly is described in Chapter 2.

Production of the Electron Beam

At the top of the TEM column is the filament assembly, which is connected to the high-voltage supply by an insulated cable (see Chapter 2). In a standard TEM, normal accelerating voltages range from 20,000 V to 100,000 V; intermediate-voltage and high-voltage TEMs may use accelerating voltages of 200,000 V to 1,000,000 V. The higher the accelerating voltage, the greater the theoretical resolution (see Table 4.1).

Below the filament tip and above the anode is a beam volume called crossover. In this area of the filament chamber, the electron beam is condensed to its highest density. There are more electrons per unit area at crossover than at any other place in the microscope. Crossover is the effective electron source for image formation. In a TEM, the diameter of the electron beam at crossover is approximately 50 μm. The anode, or positively charged plate, is below the filament assembly.

The Condenser-Lens System Controls Illumination

The electron beam next travels through the condenser-lens system. Modern TEMs have two condenser lenses. The condenser-lens system is used to control electron illumination on the specimen and on the viewing screen for such functions as viewing, focusing, and photography. How the two condenser lenses are set up in relationship to each other affects not only the amount of illumination, but also the quality of the image.

In a TEM with two condenser lenses, the first is a high-power lens, usually with a fixed current or with a limited number of current settings. On some microscopes, the variable control over the first condenser lens is sometimes referred to as the spot-size control. The first condenser lens can condense the 50-μm electron beam to as small as 1 μm. The second condenser lens is a weaker, variable lens that controls the size of the beam from 1 μm to 10 μm as it is projected to the specimen plane. The microscopist uses this lens to change the beam brightness for image viewing and photography. The current through this lens also affects the beam coherence that defines image quality.

The condenser lenses are fitted with apertures, which are usually small platinum disks or molybdenum strips with holes of various sizes. Aperture sizes in the condenser lens usually vary from 100 μm to 400 μm, depending on the requirements for specimen viewing and on user preference. Each aperture holder has a variety of sizes. Condenser-lens apertures protect the specimen from too many stray electrons, which can contribute to excessive heat and limit X-ray production farther down the column. Smaller apertures produce dimmer images but may also help protect a fragile specimen from a large dose of electrons. Apertures for condenser (and other) lenses are situated in a holder that permits the user to change and center them easily (Figure 4.7).

The most important function of a condenser lens and its aperture is to define the angular aperture of illumination. The maximum

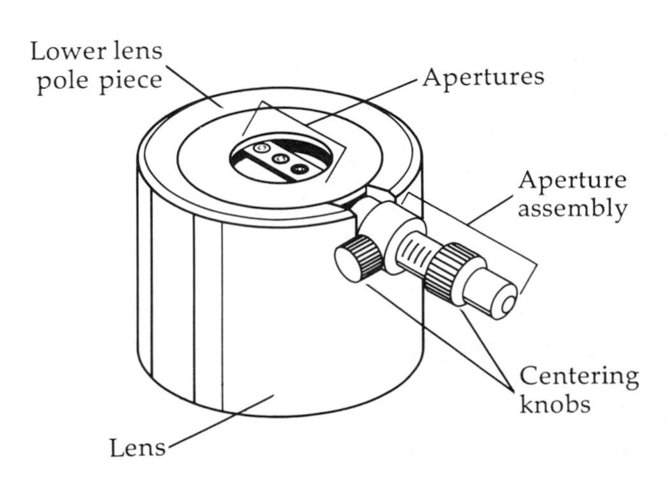

Figure 4.7 Lens aperture holder and apertures. (A) Schematic of an aperture holder and centering device. The apertures are a row of three disks. The assembly permits both X- and Y-centering of each disk in the column.

angular aperture occurs when the beam is at crossover. In this condition the beam is round and is at maximum intensity (i.e., the beam is very bright). This illuminating capacity is the highest for examining specimens but does not produce the highest image quality, especially when large condenser-lens apertures (300 μm to 400 μm) are used. Using a condenser aperture of 300 μm or less and defocusing the condenser lens decreases the angular aperture of illumination, thereby increasing beam coherence and resolution. Thus, the highest resolution occurs with defocusing of the condenser lens, but the trade-off is dimmer illumination, which can decrease the ability to focus accurately.

The Imaging Lenses Create the Magnified Image

The objective lens is the most important and complex lens in a TEM. Because it is the first magnifying lens, any imperfections in the objective lens will be magnified further by the other lenses. The specimen is inserted into the objective lens, which must be designed so that the specimen can be moved in both X and Y directions (and in most modern TEMs, also in the Z direction) and have tilting and rotating capabilities. Several designs of specimen insertion rods exist, depending on the specific microscope (Figure 4.8). Both side-entry and top-entry specimen stages exist on different TEM models. Some specimen rods hold a single specimen grid; others are designed as cartridges to hold multiple specimen grids.

The specimen (usually a 3-mm copper or other metal grid) supported in the specimen rod is inserted into the objective lens in the object plane of the lens, between the upper and lower pole pieces. The space occupied by the specimen is only a few millimeters but is enough to allow all of the movement of the specimen necessary for viewing. The

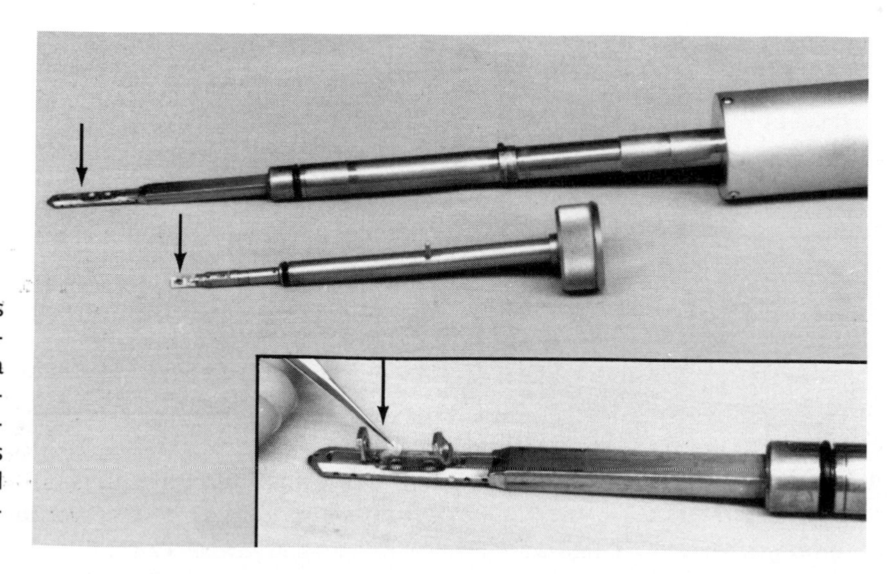

Figure 4.8 Examples of TEM specimen rods with side entry stages. The specimen is inserted at the arrow and is held in place by a retaining ring or a spring clip. Each microscope has a specimen rod designed specifically for its objective lens. The capabilities must include X, Y, and Z movement, as well as tilt and sometimes rotation of the specimen.

specimen is inserted into the microscope column through an air lock to preserve the high vacuum of the column.

Since the insertion of the specimen is a likely source of external contaminants (both from the specimen itself, which can degrade, as well as from the specimen insertion rod), most objective lenses are also fitted with a cryogenic anticontamination device (see Chapter 3), which functions as a cryopump to prevent an excess buildup of hydrocarbons on the specimen, pole pieces, and objective apertures. The anticontaminator is usually a disk or cylinder of copper or other metal that surrounds the specimen area. It has a copper braid that extends through the column to an external container (dewar) of liquid nitrogen. The liquid nitrogen keeps the metal cold to attract and hold contaminants.

Production of the Image Depends on Specific, Contrast-forming Specimen–Beam Interactions: Amplitude, Diffraction, and Phase Contrast

As the electron beam interacts with the specimen, a number of signals useful in the formation of the TEM image are generated (see Chapter 2). Several important interactions that contribute to the formation of the TEM image occur: absorption, diffraction, elastic scattering, and inelastic scattering.

Electrons may be absorbed into thick or heavily stained portions of the specimen or into areas of atoms with high atomic numbers, although very few electrons are prevented from passing through the specimen. Electrons are also absorbed into dirt or areas of contamination on the surface of the specimen. If too many electrons are absorbed into a small area of the specimen, such as a puddle of stain or particle of dirt, they may cause a buildup of heat in that area and result in distortion or even destruction of the sample. In this way, such absorption does not contribute to image formation but rather to artifacts in the image (see Further Reading). In general, absorption by the specimen does not contribute significantly to image contrast in the TEM, as opposed to the light microscope, which relies on differential absorption as the major contributor to image contrast.

Elastically scattered electrons are those produced when electrons from the beam interact with the nuclei in atoms from the specimen (Figure 4.9). The electrons undergo a large deviation in their path but little or no energy loss. Elastically scattered electrons contribute to both amplitude and diffraction contrast in an image.

Inelastically scattered electrons are those produced when electrons from the beam interact with electrons in atoms from the specimen. These electrons are characterized by a loss of energy and only a slight deviation in their path. Inelastically scattered electrons are particularly important for imaging samples of low atomic number. They can contribute to chromatic aberration because of energy loss and to phase contrast.

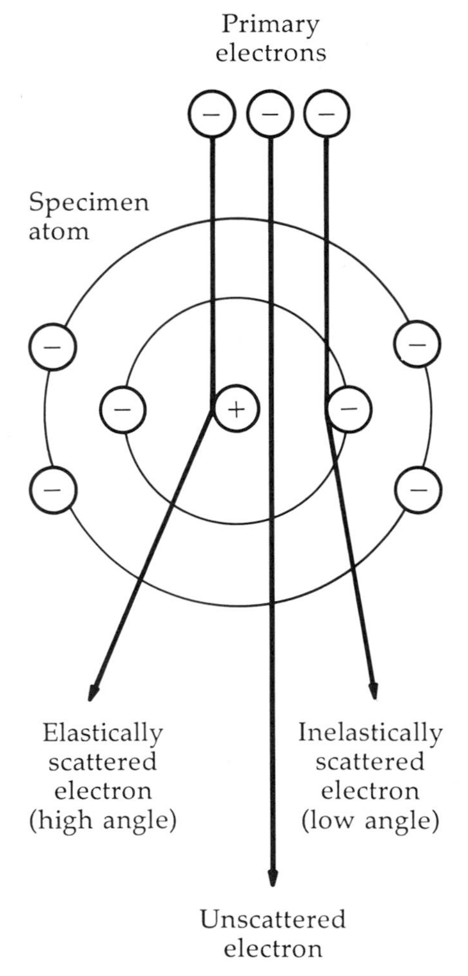

Figure 4.9 Specimen–beam interactions. Various interactions occur between the primary-beam electrons and the atoms of the specimen that contribute to the formation of the TEM image. Unscattered electrons pass through the specimen unchanged. Elastically scattered electrons are those that interact with the nuclei of atoms in the sample. These electrons are scattered through wide angles and suffer no energy loss. Inelastically scattered electrons interact with electrons of the atoms in the sample and suffer an energy loss but are scattered at low angles.

The extent to which these various interactions between electrons in the beam and atoms in the specimen result in scatter, and therefore image formation, depends on the mass–thickness of the specimen, especially for amorphous specimens (i.e., biological samples). The thickness of the specimen and the variation in thickness from one portion of the specimen to another, as well as the atomic number of the various atoms making up the specimen (the mass), have an effect on the amount of scattering that takes place. Greater thickness induces proportionally more scattering, although a specimen that is too thick causes increased chromatic aberration, which decreases resolution. Greater atomic number (mass) also results in increased scattering events. This differential scattering (called amplitude contrast or scattering absorption contrast) between the transmitted and scattered electrons from individual areas of a specimen results in the contrast necessary to form an image on the viewing screen.

In crystalline specimens in preferred orientations, scattering of the incident electrons can occur in specific directions defined by the Bragg equation: $n\lambda = 2d\sin\Theta$, where n is an integer, λ is wavelength, d is the crystal lattice spacing, and Θ is the angle of diffraction (Figure 4.10). Depending on the orientation of the crystal, scattered electrons may pass through the objective lens aperture and produce a bright spot, or most commonly, the electrons may be absorbed by the aperture, thereby producing a dark area. Diffraction contrast can be used to produce dark-field images, and when the Bragg conditions are met, crystal lattice structures may be determined.

A third type of contrast is called phase contrast. Phase contrast occurs in images of specimens that are very thin and/or have a low atomic number in which inelastically scattered electrons are not intercepted by the objective-lens aperture but rather are transmitted through it, and in which sequential, multiple events of scattering of beam elec-

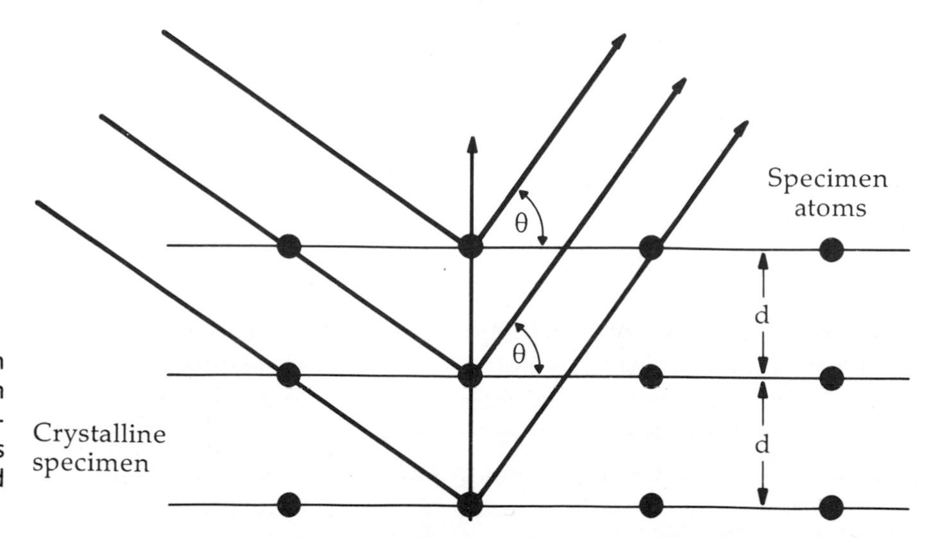

Figure 4.10 Bragg diagram of beam interaction with crystalline specimen. In crystalline samples in preferred orientations, scattering of incident electrons can occur in specific directions defined by the Bragg equation: $n\lambda = 2d\sin\Theta$.

trons do not occur. Because the inelastically scattered electrons lose energy, their wavelength changes. These electrons travel out of phase with the unscattered primary electrons and constitute the phase contrast in the image. Contrast and focus are related in the TEM image. At true focus, a very thin specimen appears transparent and flat, with little contrast. As the image is defocused, the contrast increases (a phenomenon known as defocus contrast). Thus, focus and contrast must be balanced carefully to avoid artifacts in image interpretation (see Further Reading). "Optimum underfocus" is often defined as the best compromise between contrast and focus but must be selected with care not to sacrifice the best resolution and focus for additional contrast in the image. Other options for enhancing contrast are discussed in the next section.

Two terms used to describe these variations in contrast (i.e., variations in scattering) are electron-transparent (or electron-lucent) and electron-dense. "Electron-transparent" refers to areas that scatter fewer electrons and appear as bright areas in the image. "Electron-dense" refers to areas that scatter more electrons, as well as absorb electrons (although these are few), and appear as dark areas in the image. The contrast, then, is due to the formation of discrete electron-transparent and electron-dense areas in the image.

Enhancing Contrast Entails Balancing Trade-offs

The objective lens in a TEM, like the condenser lens, is equipped with a set of apertures (see Figure 4.7). These may be platinum disks, molybdenum strips, or special gold-foil disks. The sizes of the holes usually range from 20 μm to 50 μm or larger, depending on the specimen and the needs of the operator. The smaller apertures become contaminated more quickly than the larger apertures. Gold-foil apertures actually heat up when exposed to the beam and are, therefore, considered to be "self-cleaning". The objective apertures are located between the specimen and the lower pole piece of the lens.

The objective apertures in a TEM are also referred to as contrast apertures. They absorb most of the elastically scattered electrons and a few of the inelastically scattered electrons from the image, resulting in a subtractive effect (Figure 4.11). The smaller the objective aperture, the

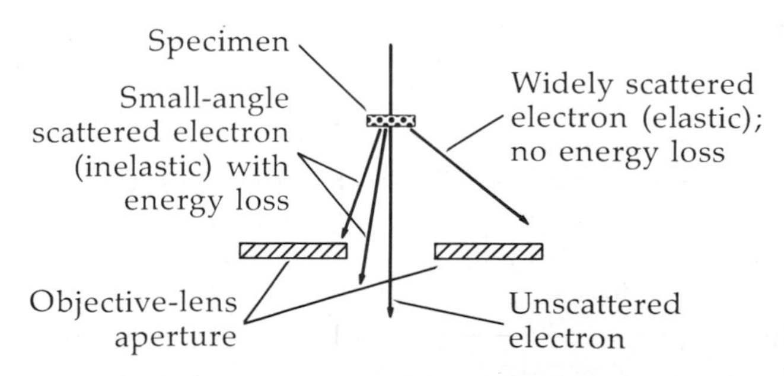

Figure 4.11 Enhancing contrast with the objective aperture. The objective-lens aperture, also called the contrast aperture, allows unscattered and most inelastically scattered electrons to continue down the column to strike the screen but does not permit electrons scattered through a large angle (elastic electrons) to continue. The aperture thus imparts contrast by differentially allowing some electrons to pass through and not others.

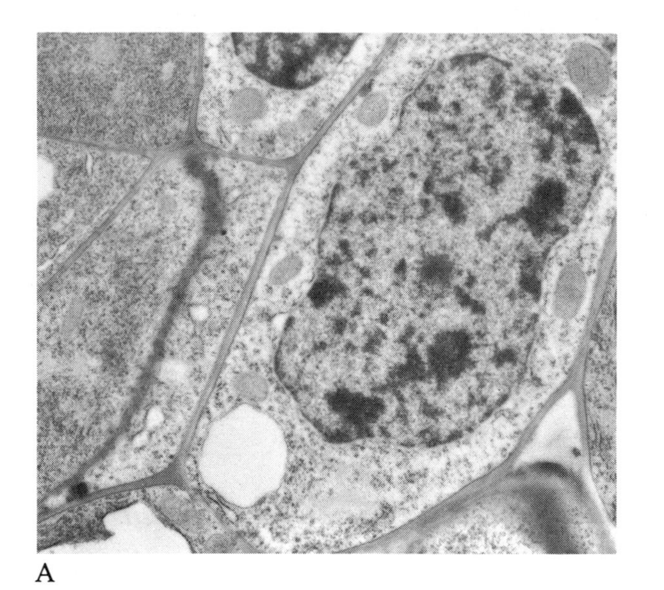

A

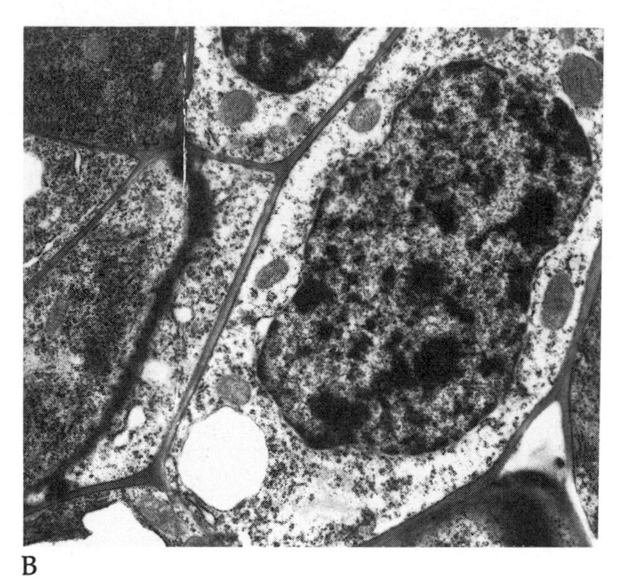

B

Figure 4.12 Contrast effects of the objective aperture. (A) An ultrathin section of a plant photographed with no objective aperture. (B) An ultrathin section of the same sample photographed with an objective aperture of 40 μm. Use of an objective aperture greatly enhances amplitude contrast of the image because more elastically scattered electrons are excluded.

greater the contrast, because more scattered electrons are removed from the image (Figure 4.12). A small objective aperture decreases the spherical aberration, which is a major limiting factor for resolution, but because a small aperture also decreases the numerical aperture of the lens, it also decreases resolution.

As with many operations in electron microscopy, a compromise must be reached in trying to gain the advantage of increased contrast. The smaller objective apertures may cut off a portion of the image, particularly at lower magnifications, and become contaminated more quickly, even if the anticontaminator is used. A contaminated aperture usually results in objective-lens astigmatism (see next section and Chapter 2).

In addition, decreasing the accelerating voltage results in an increase in contrast because the electrons move through the specimen at a lower energy, therefore interacting with more atoms of the specimen. Again, a compromise must be made, since decreasing the accelerating voltage decreases resolution and requires thinner specimens to create a high-quality image, although it increases amplitude contrast. Chromatic aberration and radiation damage to the specimen (discussed later in this chapter) also increase with decreasing accelerating voltage.

Objective-Lens Astigmatism Decreases Resolution

Objective-lens astigmatism is a common, yet correctable, problem in a TEM. As already pointed out, it is most often caused by contamination on the objective aperture. Chapter 2 explains the use of the octapole stigmator, which can be used to correct astigmatism. It is important to remember that objective astigmatism changes as the objective lens apertures and/or specimens are changed and becomes more obvious at higher magnifications. Objective-lens astigmatism results in micrographs of poor quality (Figure 4.13A). An image of good quality (Figure 4.13B) requires that astigmatism be corrected frequently, and at the minimum, that the image be checked for evidence of astigmatism before every micrograph is taken.

Objective-lens astigmatism can be corrected using the specimen image itself or using a special specimen called a holey film. The holey film is a thin plastic film with numerous small holes that is applied to a copper grid and stabilized with gold or carbon. The holes form opaque edges with which the the electron beam can interact to produce the diffraction effects of Fresnel fringes (Figure 4.14). Since the Fresnel fringes are very sensitive indicators of the quality of the image, they can be used to diagnose objective lens astigmatism and to correct it.

If the TEM objective lens is astigmatic, the Fresnel fringes will be at differing distances and locations within the hole. For example, at two opposite points the fringe may be outside the hole, and at two opposite points 90° to them, the fringe may be inside the hole. The position of the fringes can be changed using the stigmator correction controls and the fine-focus controls to correct the astigmatism by making the fringe even

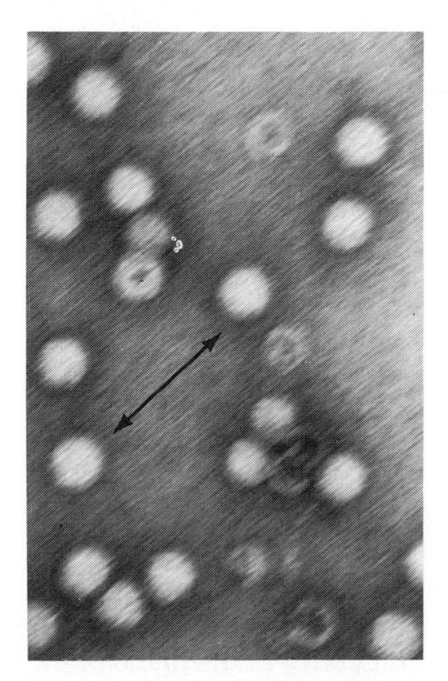

A

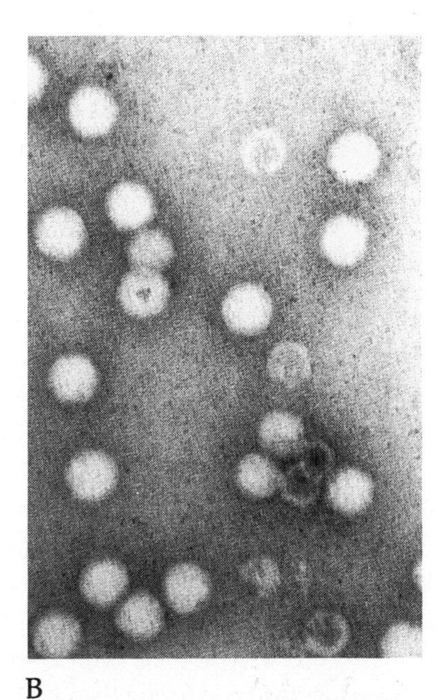

B

Figure 4.13 Effects of objective-lens astigmatism. (A) This image of negatively stained virus particles is well focused but astigmatic. Note the blurring or pulling in the direction of the arrow, which indicates objective-lens astigmatism. Changing the focus emphasizes the pulling, making it easier to diagnose astigmatism. This image is obviously of very poor quality but can be corrected by use of the objective-lens stigmators. (B) The image has been corrected for astigmatism. The pulling is gone. Each image should be examined for astigmatism before being photographed. Judicious attention to astigmatism and proper correction of this defect will result in improved micrographs.

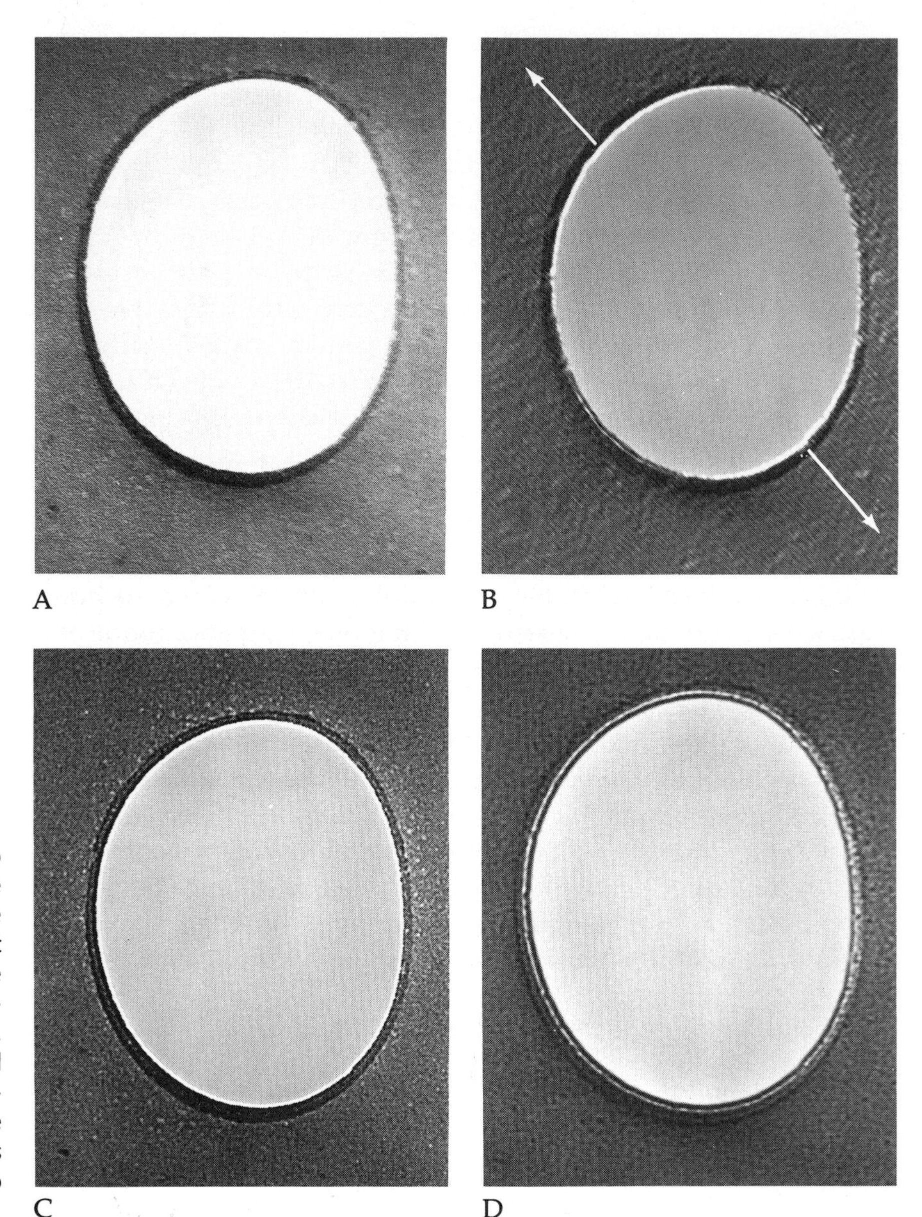

Figure 4.14 Use of the Fresnel fringes to examine focus and astigmatism in the TEM image. By careful attention to the location of the Fresnel fringe (the bright line associated with the hole), it is possible to determine (A) a true-focus and nonastigmatic image, (B) an astigmatic image, (C) an underfocused image (Fresnel fringe inside the hole), and (D) an overfocused image (Fresnel fringe outside the hole). Arrows indicate the Fresnel fringes used to correct astigmatism and also to determine the condition of focus.

around either the inside or outside of the hole. The graininess and appearance of the plastic film gives a good indication of the effects that astigmatism has on a final image (see Figure 4.13A).

The position of the Fresnel fringe can also provide information about whether a TEM image is at under-, over-, or true focus (Figure 4.14). If the TEM objective lens is overfocused (i.e., too much current is passing through it), the Fresnel fringe will be located on the outside of the hole. If the lens is underfocused, the fringe will be on the inside. If no fringe is visible, the lens is at true focus, exactly in the plane of the hole edge. At true focus the contrast in a very thin specimen is minimal.

Formation of a Final Magnified Image Depends on All of the Imaging Lenses

Final image magnification in a TEM is the product of the magnifications of each magnifying lens: the objective lens, the diffraction lens, the intermediate lens, and the projector lens (see Figure 4.6). The objective lens, the first magnifying lens in the TEM, is a high-power lens. The diffraction lens, below the objective lens, is a very low-power lens. The intermediate lens, below the diffraction lens, is a weak lens, but with variable power. It is the current in this lens that is adjusted to control the final magnification of the image. The last lens is the projector lens, which is a high-power lens that projects the final magnified image onto the viewing screen.

A TEM image must be focused after every magnification change. Changes in magnification are accomplished by changing the current in the intermediate lens. Hence, the magnification control on the TEM panel is actually an intermediate-lens current control. When the intermediate-lens current is adjusted, both the final magnification of the image and the object plane of the intermediate lens change. Focusing, accomplished with the objective-lens controls, involves altering the current in the objective lens so that its image plane is kept coincident with the object plane of the intermediate lens (which was altered with the magnification change) (Figure 4.15).

As with the correction of astigmatism, Fresnel fringes can be used to learn how to focus and to recognize under-, over- and true-focus images. If the TEM objective lens is overfocused (i.e., too much current is passing through it), the Fresnel fringe will be located on the outside of the hole (see Figure 4.14). If the lens is underfocused, the fringe will be on the inside of the hole. If no fringe is visible, the lens is at true focus (i.e., it is exactly in the plane of the hole edge). The image itself can also

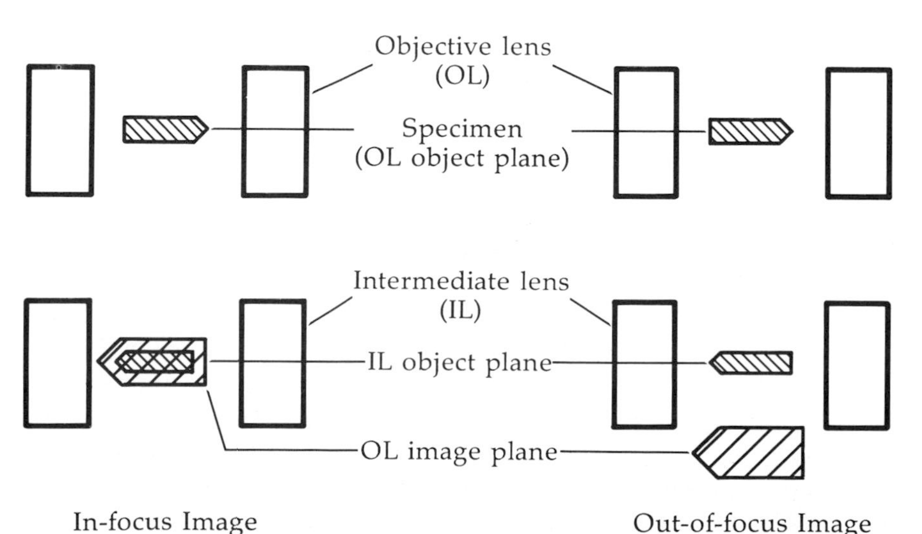

Figure 4.15 In-focus and out-of-focus setup in a TEM. The intermediate lens controls the final magnification of the image. After every change in magnification using the intermediate-lens control, the image must be focused using the objective-lens control. The image plane of the objective lens must be coincident with the object plane of the intermediate lens for the image to be in focus.

be examined at each of these focus conditions to ascertain the effect of focus on it.

The final image of the specimen is projected onto the viewing screen by the projector lens (see Figure 4.6). The screen is in an observation chamber equipped with one or more windows of leaded glass (Figure 4.6). The screen is coated with a fluorescent powder of zinc and cadmium sulfide. It emits visible light when struck by the electrons carrying the image information, thus making it possible to view the image. The zinc–cadmium sulfide is a fine-grain powder but is still grainier than the emulsion on the photographic film. Thus, information in the image may be resolved, or at least resolved better, only on a micrograph, not on the screen itself. The zinc–cadmium sulfide powder is prepared as a yellow-green, the color to which the dark-adapted eye is most sensitive.

Recording the Image for Future Use

The TEM may be fitted with any number of cameras or other devices to record images. The great depth of focus of a TEM allows placement of the camera(s) either above or below the viewing screen. A plate camera, and sometimes a 35-mm camera, is usually available. A video camera may also be attached to store images on a videotape for further processing by computer or for comparing many images at a later date. The photographic process is explained in Chapter 9.

Since the depth of focus in a TEM is very large (see Figure 4.4B), the camera(s) may be placed in a variety of locations below the projector lens. With the cameras placed anywhere within an area from just below the projector lens to the area below the viewing screen, the image remains in focus and can be recorded.

Most microscope manufacturers guarantee their magnification display readouts to be within 10% of the actual value. In practice, most magnification readouts are well within this limit. Variations in lens currents used for magnification and focusing, as well as the exact position of the specimen, can contribute to a magnification readout that is slightly more or less than the actual value. When very accurate magnifications are required, it is necessary to use either an internal standard, such as latex spheres of known size or a magnification calibration standard with known spacings mounted on a microscope grid (Figure 4.16). Standards should be used on a routine basis to check the accuracy of the magnification readouts.

Figure 4.16 Magnification calibration standard. A shadowed sample of a waffle-type diffraction grating can be used to calibrate magnification in the TEM. The diffraction grating has spacings of known size. When photographed using the same microscope focus settings as were used with the specimen of interest, the grating can be used to compute magnification accurately. The grating shown here has spacings of 2,160 lines/mm.

An internal magnification calibration standard is photographed along with the specimen. The magnification standard on a separate grid is photographed immediately after the completion of micrographs of the specimen. The standard is photographed at each of the magnification steps of the specimen micrographs for which a more accurate magnification is needed. Care should be taken not to alter the objective lens (focus), or to do so as little as possible. Keep in mind that accurate measurements of the magnification calibration negatives or contact prints are necessary to minimize the degree of error caused by sloppy use of a ruler.

Selected-Area Diffraction

Along with the usual TEM image of a specimen, a diffraction pattern is formed at the focal point of the back focal plane (see Figure 4.3) of each imaging lens. By altering the current in the diffraction lens, one can view this image in the back focal plane of the objective lens instead of the usual bright-field image. This diffraction pattern is formed by the elastic scattering of the electron beam by the atoms in the specimen (Figure 4.17). Such a pattern follows the demonstration by William Bragg that crystal structure can be determined by X rays according to a specific mathematical formula. As mentioned earlier, the Bragg equation states: $n\lambda = 2d\sin\Theta$, where d is the distance between planes of the crystal, λ is the wavelength of the energy source used, n is the order of the spectrum, and Θ is the angle of incidence (or reflection) (see Figure 4.10). Bragg received a Nobel Prize in Physics in 1915 for this work.

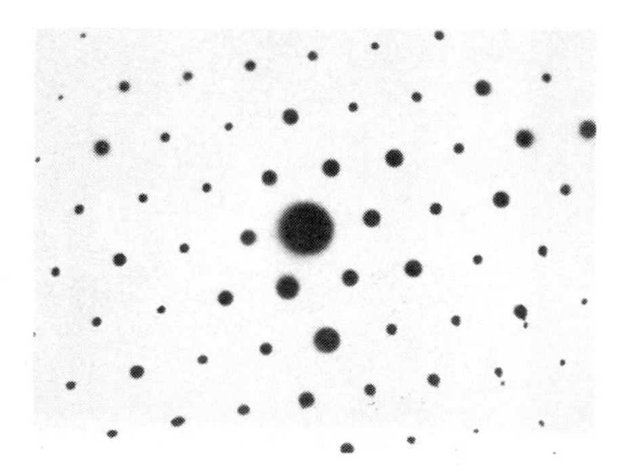

Figure 4.17 Example of electron diffraction. A mica sample placed on a carbon-coated grid shows a sharp electron diffraction pattern with spots demonstrating crystallinity.

Briefly, if an area of a specimen is crystalline in nature, the scattering occurs in a regular manner, giving a pattern of sharp spots. If an area is polycrystalline, the spots are distributed to produce sharp lines. If an area is amorphous, as with most biological specimens, the pattern is one of very diffuse rings.

Selected-area diffraction may be used to determine areas of ordered or crystalline nature in a biological specimen, such as mineral deposits or asbestos fibers. The pattern of spots or sharp rings contains information about the distances between atoms and thus can be used to determine the arrangement of the atoms. The specific orientation of a crystalline area in the specimen, with respect to the electron beam, determines the pattern and therefore the ease of interpreting the pattern. Specific areas of a specimen may be chosen for examination by selecting a particular diffraction aperture size to define a given area of the specimen (hence the name, selected-area diffraction). A much more detailed description of electron diffraction is covered in Beeston et al. (1972) in Further Reading).

Dark-Field Imaging

A dark-field image can be formed by using only the widely scattered electrons from the specimen (Figure 4.18). The unscattered electrons and those inelastically scattered electrons that are not absorbed by the objective-lens aperture are blocked from contributing to the dark-field image. The dark-field image has high contrast, although the image itself may be dimmer than a standard bright-field image.

A

B

Figure 4.18 Comparison of bright-field and dark-field images. (A) A bright-field image of a mica sample on a plastic and carbon-coated grid. (B) A dark-field image of the same mica sample. Dark-field imaging uses the elastically scattered electrons that are not used in the standard bright-field image.

Blocking the unscattered electrons and those scattered through only small angles can be accomplished quickly by off-setting the objective aperture to allow only the highly scattered electrons to form the image. For high-resolution dark-field imaging, however, using the circuit on the TEM that can tilt the beam electronically results in an image of much higher quality.

Electron Energy Loss Spectroscopy

Electron energy loss spectroscopy (EELS) is both an analytical and an imaging mode for use on a TEM or on an STEM (see Chapter 5). The method is based on inelastic scattering of the primary electron beam by the electrons of the atoms in the sample. This scattering causes a loss in primary-electron energy; thus, the EELS spectrum is an energy distribution of electrons transmitted through a thin specimen.

The EELS detector is placed in the microscope so that it can intercept and disperse the inelastically scattered electrons as they are transmitted through the specimen. All electrons of the same energy are focused at the same point on the detector within a defined cone of acceptance. In this way, the detector acts as a lens, and because it uses a magnetic field to focus the electrons, it is often called a magnetic prism. Because the EELS data are collected sequentially, each channel in the spectrum may contain only a few counts. This is a limitation of the EELS method. Newer methods collect spectra in parallel (PEELS) to minimize this problem.

The EELS method is very sensitive to elements of low atomic number. It has higher spatial resolution than energy-dispersive X-ray analysis (see Chapter 8) because only the electrons within the cone of acceptance are detected. Increasing the thickness of the sample alters the EELS spectrum in two important ways: (1) the energy loss peaks become less sharp and (2) multiple scattering can produce extra peaks. Since interpretation of the spectrum is based on single scattering events, the effects of specimen thickness must be considered to ensure accuracy. Compositional imaging can also be done using EELS by digitizing the data and then displaying those portions of the spectrum associated with the elemental peaks of interest. Electron energy loss is used predominantly in materials sciences but can also be used to quantitate the thickness of sections in biological sciences.

MEDIUM- AND HIGH-VOLTAGE TRANSMISSION ELECTRON MICROSCOPY

Conventional TEMs operate at accelerating voltages of between 60,000 V and 120,000 V. There are, however, high-voltage TEMs that operate at a million volts or more and medium-voltage TEMs that operate at 200,000 V, 300,000 V, and 400,000 V. High-voltage TEMs require large, heavy, high-voltage tanks and very strong lenses. The instruments generally are housed in special rooms of two to two-and-a-half stories. The medium-voltage instruments are becoming more popular both for materials science specimens and for some biological applications. These instruments are less costly, require much less space, and provide almost all of the advantages of a high-voltage TEM.

With a medium- or high-voltage TEM, thicker specimens can be examined. Specimen thickness in a conventional-voltage TEM is limited by chromatic aberration to approximately 100 nm to 120 nm. In a one million–volt TEM, a 1-μm specimen can be examined, and if the specimen is open or porous, as biological cells or tissues are, even thicker samples can be viewed. The increase in accelerating voltage allows for thicker specimens to be penetrated, making three-dimensional reconstructions of biological cells and tissues less tedious than reconstructions of the serial ultrathin sections that are necessary with conventional accelerating voltages.

An additional consideration is that with very thin specimens, especially biological specimens of inherently low contrast, the overall contrast of the image is reduced because of the smaller number of scattering events. Thicker specimens result in increased contrast decreased resolution. The increased accelerating voltage assists in optimizing the balance between specimen thickness and resolution. For some specimens, most notably materials-science specimens, increased accelerating voltage combined with a very thin specimen allows for the highest possible resolution.

With conventional TEMs, considerable beam damage takes place with organic specimens because the electrons in the beam displace and/or ionize specimen atoms. With medium- and high-voltage TEMs, there are fewer interactions between the beam electrons and the specimen atoms because of the increased accelerating voltage. Because the ionization probability is decreased with increased accelerating voltage, there is usually less beam damage. The reduction of beam damage is not linear, however, and is reduced by a factor of three when the accelerating voltage is increased from 100 kV to 1,000 kV.

In summary, the medium- and high-voltage TEMs have two primary advantages: increased resolution and the capability to image thick specimens. Although the cost of the instruments is higher than that of conventional TEMs, the medium-voltage TEMs in particular can provide useful imaging capabilities for both biological and materials sciences.

FURTHER READING

Agar, A. W., R. H. Alderson, and D. Chescoe. 1974. *Principles and Practice of Electron Microscope Operation.* North-Holland, Amsterdam, and Elsevier, New York.

Beeston, B. E. P., R. W. Horne, and R. Markham. 1972. *Electron Diffraction and Optical Driffraction Techniques.* North-Holland, Amsterdam, and Elsevier, New York.

Cosslett, V. E. 1951. *Practical Electron Microscopy.* Academic Press, New York. (Cosslett's rule)

Crang, R. F. E., and K. L. Klomparens. 1988. *Artifacts in Biological Electron Microscopy.* Plenum, New York. (Artifacts in image interpretation, setting up microscope variables)

Joy, D. C., A. D. Romig, and J. I. Goldstein, eds. 1986. *Principles of Analytical Electron Microscopy.* Plenum, New York. (Electron energy loss spectroscopy)

Leapman, R. D., and J. A. Hunt. 1992. Compositional imaging with electron energy loss spectroscopy. In *Microscopy: The Key Research Tool.* (A special publication of the Electron Microscopy Society of America.) EMSA, Woods Hole, MA.

Meek, G. A. 1976. *Practical Electron Microscopy for Biologists.* Second edition. Wiley, New York.

Misell, D. L., and E. B. Brown. 1987. *Electron Diffraction: An Introduction for Biologists.* North-Holland, New York. (Selected-area diffraction)

Reimer, L. 1984. *Transmission Electron Microscopy. Physics of Image Formation and Microanalysis.* Springer-Verlag, Berlin.

5

The Scanning Electron Microscope

The scanning electron microscope (SEM), first introduced to the commercial market in 1965, has proven to be an extremely useful scientific instrument, although its initial potential was greatly underrated. A survey done in 1963 predicted that approximately ten instruments would be the maximum number number to be sold. In the first decade, however, thousands of SEMs were sold, and their sales rate continued to increase in the second decade. One manufacturer's model had a total production run in excess of fourteen hundred machines.

There are various reasons for the popularity and usefulness of the SEM. The SEM has a large depth of field (i.e., the amount of the sample that can be in sharp focus at one time), which can be up to four hundred times greater than that of a light microscope. Much of the analysis of SEM samples is done within the magnification range of the light microscope. The total information content of the image from an SEM, can be much greater because of its great depth of field (Figure 5.1). The SEM also has higher resolution than that of a light microscope; thus, samples can be examined at a higher magnification than is possible with a light microscope (Figure 5.2). The instrument has an extremely wide range

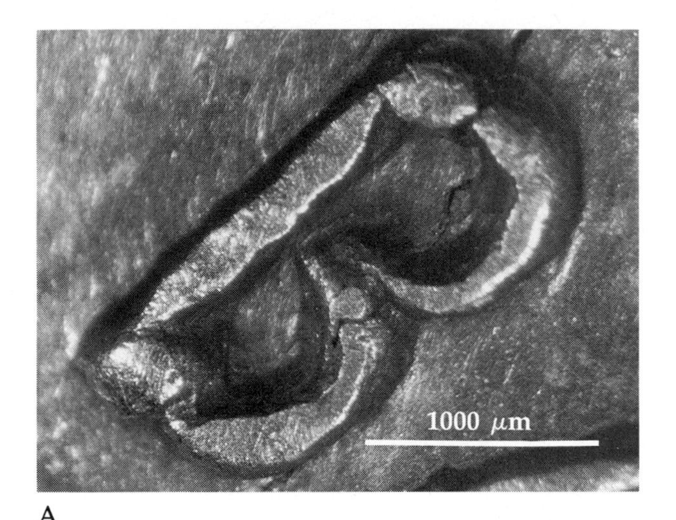

A

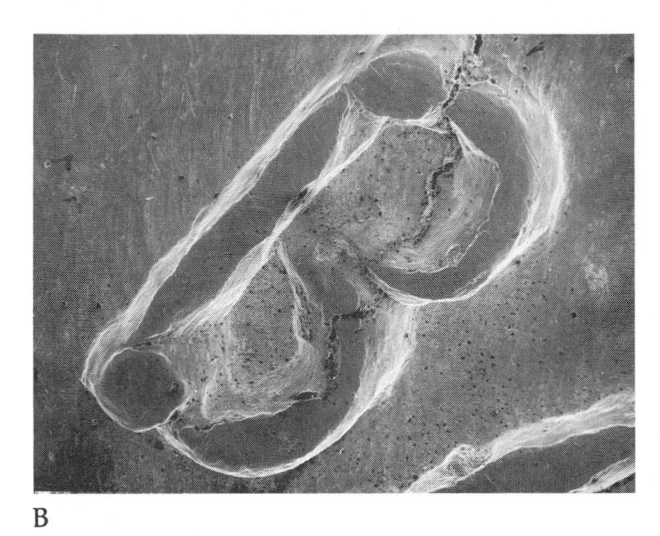

B

Figure 5.1 Comparison of SEM and light-microscope images at low magnification. (A) The letter ß on an ancient Greek coin of Ptolemy I from 300 B.C. The coin was photographed using a high-quality light microscope. The top of the letter is in sharp focus; the bottom of the letter and the background are not in focus. Depth-of-field is a limiting factor. (B) The same coin photographed in an SEM. All features are now in sharp focus; depth-of-field is no longer a limiting factor.

of magnification, usually between about 10× and 100,000×. Most techniques for the preparation of samples for the SEM are considerably easier than for the TEM because the surface of whole samples is examined, and no sectioning is required. All of these advantages have made the SEM the instrument of choice for a variety of applications.

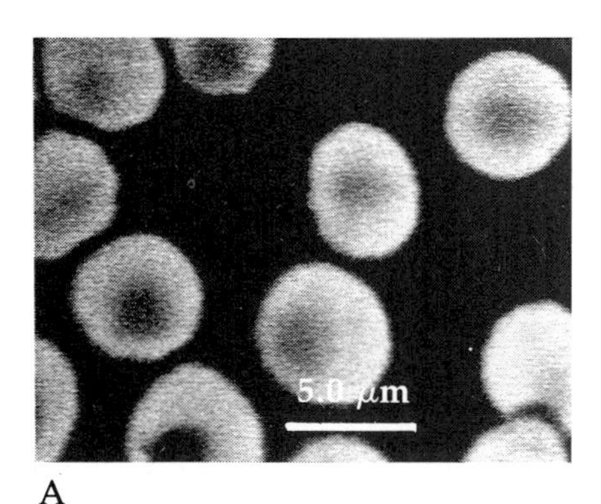

A

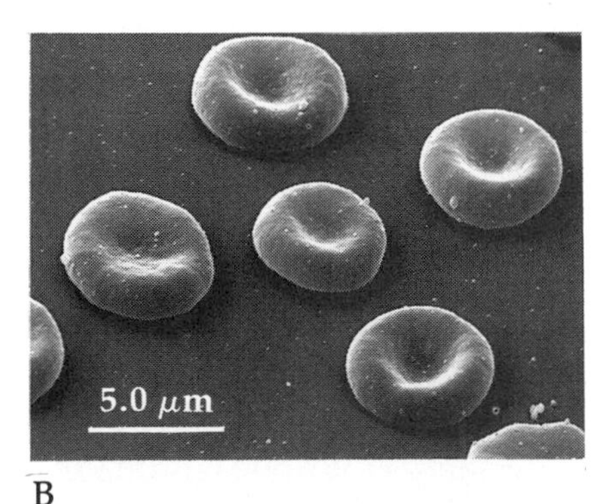

B

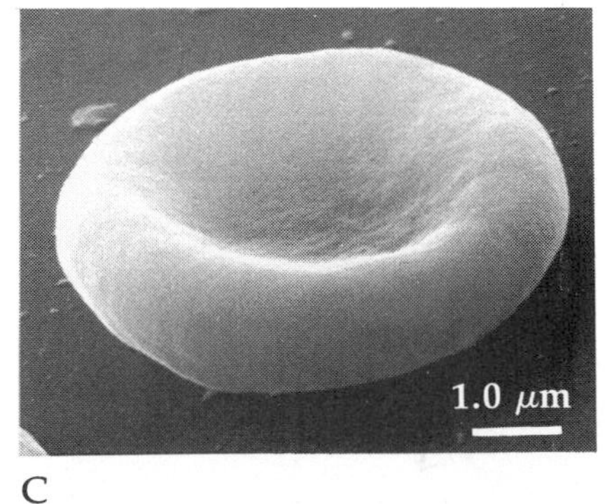

C

Figure 5.2 Comparison of SEM and light-microscope images at medium to high magnification. (A) Human red blood cells photographed in a laser scanning microscope at the upper limit of magnification for light microscopes, 4,000×. Resolution is very limiting, and most details of the cells are not clear. Photo courtesy of J. H. Whallon. (B) Red blood cells photographed in an SEM at the same magnification. The superior resolution of the SEM produces a much sharper image. In addition, it is possible to view the cells tilted to give a perspective view. (C) A red blood cell photographed by an SEM at a magnification beyond the range of the light microscope. Good resolution is still evident in the image.

THEORY OF OPERATION

Anatomy of the SEM

A basic diagram of an SEM is shown in Figure 5.3. The electron gun produces a beam of electrons that is attracted through the anode and condensed by the condenser lens and then focused as a very fine point on the specimen by the objective lens. A set of small coils of wire, called the scan coils, is located within the objective lens. The coils are energized by a varying voltage produced by the scan generator and create a magnetic field that deflects the beam of electrons back and forth in a controlled pattern called a raster. The raster is very similar to the raster in a television receiver.

The varying voltage from the scan generator is also applied to a set of deflection coils around the neck of a cathode-ray tube (CRT). The magnetic field from this coil causes the deflection of a spot of light back and forth on the surface of the CRT. The pattern of deflection of the beam of electrons on the sample is exactly the same as the pattern of deflection of the spot of light on the CRT.

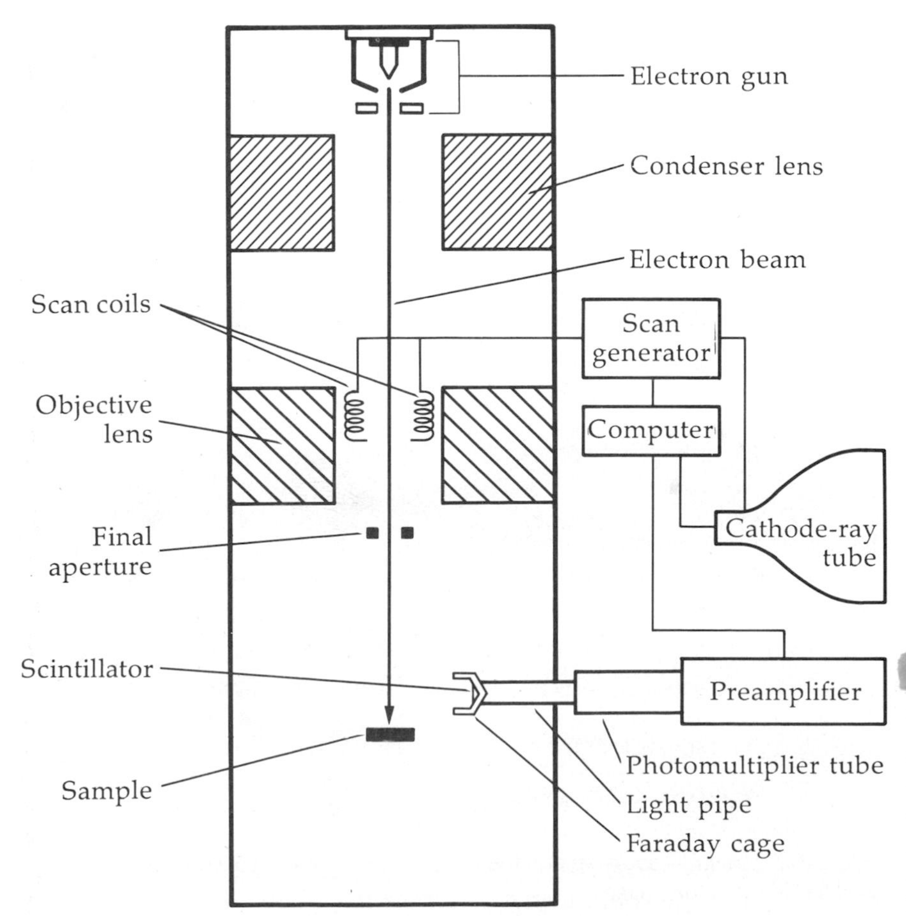

Figure 5.3 Schematic of a scanning electron microscope. The electron beam scans the sample while a spot scans the surface of a CRT. No paths of rays connect the image with points on the sample as in a light microscope or TEM. The image is thus not a true image but instead a map of the surface.

When the beam of electrons strikes the sample, a complex series of interactions occurs, resulting in the production of secondary electrons from the sample, which are collected by the detector, converted to a voltage, and amplified. The amplified voltage is then applied to the grid of the CRT and modulates or changes the intensity of the spot of light on the surface. For example, if at a given instant the beam is on a projection on the surface, a large number of secondary electrons will be detected, causing a large voltage in the detector that results in a bright spot on the surface of the CRT. If the beam of electrons then moves to a depression on the sample, fewer electrons will be detected and a smaller voltage will be developed in the detector, resulting in a darker spot on the surface of the CRT. The SEM image, then, consists of thousands of spots of varying intensity on the face of a CRT that correspond to the topography of the sample.

Everhart-Thornley Detector

The Everhart-Thornley detector is a key component in the SEM (Figure 5.4). The front of the detector has a structure called a Faraday cage or collector screen, which is either a wire-mesh or a metal ring with a positive 300-V electrical charge that attracts the low-energy secondary electrons. Once inside the Faraday cage, the secondary electrons are accelerated by the 12,000-V positive electrical charge applied to a scintillator, which is a metal-coated disk that acts as a collector. When the secondary electrons strike the scintillator, they are converted into a photon of light. The light impulses are then transmitted through the light pipe, which terminates outside of the specimen chamber (see Figure 5.3).

The photon exits the light pipe and then enters a photomultiplier tube, where it strikes the first electrode, causing it to emit electrons. The electrons then bounce back and forth among a series of charged plates inside the tube, and with each bounce the number of electrons is multiplied in a cascading manner, ultimately producing a gain of 10^6 electrons. The output voltage of the photomultiplier is then amplified further by the preamplifier. The preamplifier output voltage ultimately modulates the intensity of the spot on the cathode-ray tube.

The Everhart-Thornley detector detects nearly all of the secondary electrons that escape from the surface of the specimen because they are low in energy and can be deflected easily through very large angles. The detector also detects backscattered electrons, which are already traveling directly toward the scintillator. Because of their high energy, backscattered electrons are not easily deflected.

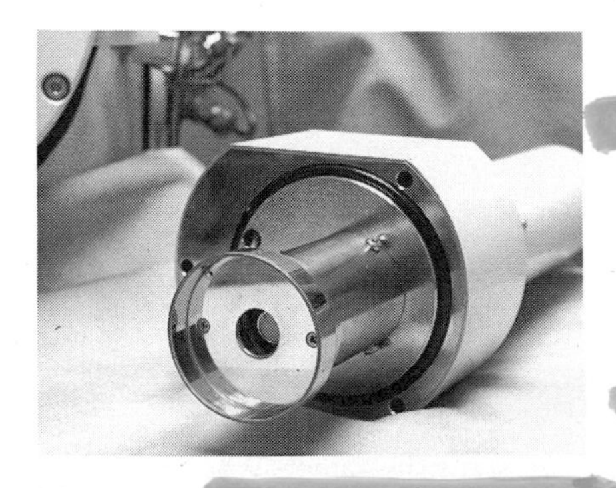

Figure 5.4 Everhart-Thornley detector. This photograph of an Everhart-Thornley detector shows the Faraday cage, the ring-shaped structure at the front. The scintillator is the small disk just inside the ring.

The Faraday cage is necessary to protect the beam of electrons from the 12,000-V positive electrical charge on the scintillator. Without it, the charge could deflect the electron beam or cause distortion in the beam. The Faraday cage confines the action of the 12,000-V force field, and because it is external to it, the beam of electrons is influenced only by the 300-V force field of the Faraday cage itself.

The scintillator is a small disk of about 8 mm to 20 mm in diameter whose function, as previously mentioned, is to convert the secondary and backscattered electrons into a burst of light. Originally, scintillators were made of a special type of light-emitting plastic. Most SEMs now use scintillators coated with a phosphor that produces light under electron bombardment. All scintillators have to be coated with a very thin coating of aluminum to make them electrically conductive to carry the 12-kV force field. The aluminum does not prevent the passage of the secondary and backscattered electrons into the scintillator.

Image Recording

Images from the SEM are produced on a cathode-ray tube (CRT). Most SEMs employ two types of CRTs, one for viewing and one for photography. The viewing CRTs have a persistent phosphor that glows for several seconds after the scan moves across the screen. Since the image persists, it is easy to view. Unfortunately, the persistent phosphors by nature are very coarse-grained and give poor resolution and, therefore, poor image formation. For this reason the photography is done with a second CRT (called the record CRT) that has a fine grain, a phosphor of short persistence, and thus very high resolution. Most manufacturers now use record CRTs with a 2,500-line resolution capacity. The image on the record CRT is hard to view because only the scanning lines are present, and no image persists. A camera is mounted in front of the CRT, and the image is recorded line by line on film, which does not require persistence of the image. Many laboratories use instant film to know immediately if they have obtained an acceptable image. Adapters are available, however, that allow the use of large-format roll film or 35-mm film.

Digital SEMs Have Many Advantages over Analog SEMs

Nearly all SEMs now being built are digital rather than analog. In addition, many manufacturers offer digital conversions for analog SEMs. Unlike with analog SEMs, in digital SEMs a computer controls scan generation and the processing of signals from the detector. The scan and image are divided into a matrix of small points (called pixels for picture elements) rather than lines. Common formats are 512 by 512 and 1,024 by 1,024 pixels.

Digital SEMs have several major advantages over the older analog versions. One major advantage is that of frame-averaging of the image. Frame-averaging allows one to improve a very low-quality, "noisy" image with a poor signal-to-noise ratio to produce a crisp clear image (Figure 5.5). Frame-averaging may be accomplished by one of two

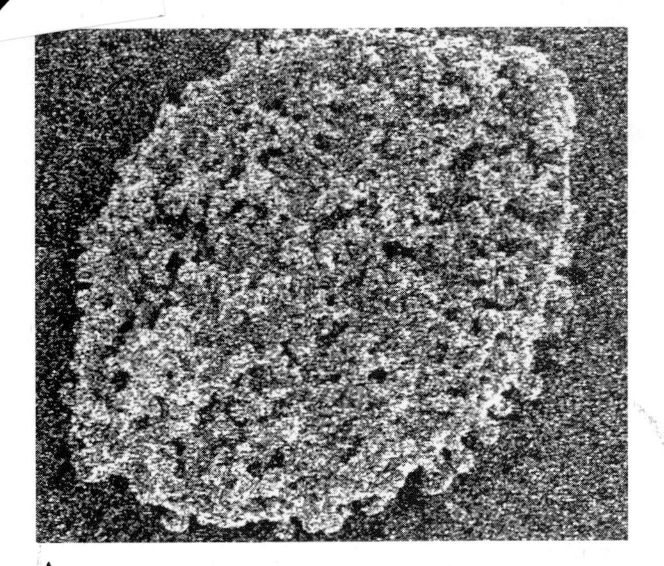

A

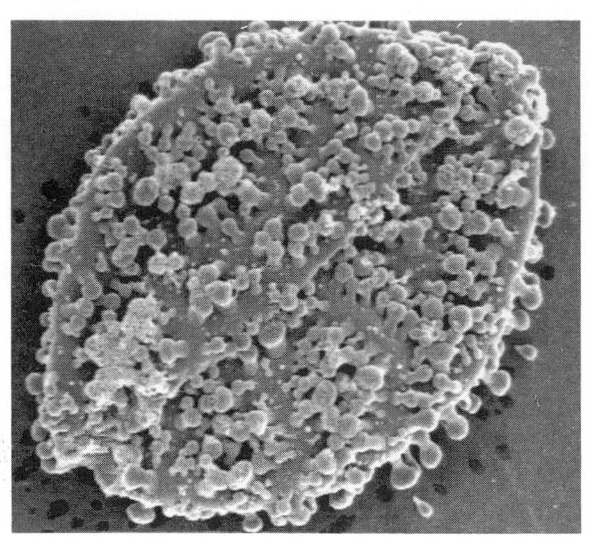

B

Figure 5.5 Frame-averaging in a digital SEM. (A) A diatom imaged with two frames averaged. The image is very poor because of the electronic noise present. (B) The same sample with one thousand frames averaged. The image is now much clearer.

methods: pixel-averaging or true frame-averaging. In pixel-averaging the beam dwells on each pixel longer, the amount of time being adjustable. By dwelling longer, a better signal-to-noise ratio may be obtained. True frame-averaging uses a fast television-rate scan. The first frame image is put into a buffer memory; then another frame is scanned and added, and finally, all the frame images are averaged. Usually 100 to 200 frames are collected; however, up to 2,000 are often possible.

Another advantage of digital SEMs is that the image may be stored in memory, rather than on film, resulting in a considerable savings in film costs. A 1,024-by-1,024 image requires approximately one megabyte of memory. Thus, one image can be stored on a 1.4-megabyte floppy disk, four to 16 in temporary RAM memory, and 40 to several hundred on a hard drive, depending on the disk size. Images may be retrieved from memory at any time and photographs taken as needed.

Digital SEMs also have reduced charging (see Chapter 7). Fast television-scan rates reduce charging problems because the charge does not have time to build up, since the time that the electron beam dwells on the image is short. Television-scan rates normally give poor resolution. Because of the pixel-averaging ability in a digital SEM, however, television-scan rates do not result in a loss of resolution.

One final advantage of digital SEMs is that an interface to high-resolution printers with the 1,024-by-1,024 format is possible. Often such images are adequate for documentation in place of using film. The cost of images from a printer are a small fraction of the cost of photographic film.

One disadvantage of digital SEMs is that the resolution of the photo is reduced. A photo taken with the 1,024-by-1,024 pixel format on a digital SEM has less resolution than a photo taken on a 2,500-line

CRT in an analog SEM. The difference, however, is noticeable only in an enlargement made from a negative.

SPECIMEN–BEAM INTERACTIONS

When an incident-beam electron strikes the surface of a sample, it undergoes a series of complex interactions with the nuclei and electrons of the atoms of the sample. The interactions produce a variety of secondary products, such as electrons of different energy, X rays, heat, and light. Many of these secondary products are used to produce the images of the sample and to collect additional data from the sample.

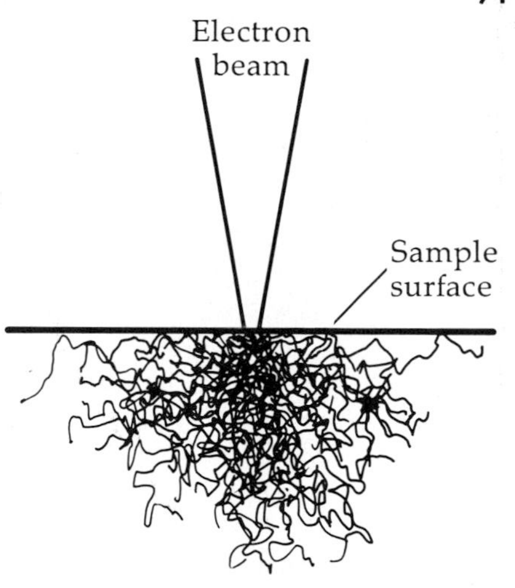

Figure 5.6 Scattering of beam electrons in the interior of a sample. Individual electrons may have different scattering angles and path lengths as they interact with the atoms of the sample.

Interaction Volume

When a beam electron interacts with the atoms of the sample, each interaction causes the incident electron to change direction, and many interactions cause it gradually to lose energy. The interaction is a scattering process in which there are no sharply defined limits to the extent of scattering (Figure 5.6). For ease of illustration, a defined limit may be placed on the remaining incident electron energy and on the region of interaction illustrated. Such an illustration would describe the area in which the interactions are most likely to occur. The interaction is usually described as teardrop- or pear-shaped. In whole samples, the volume (both depth and width) of the interaction varies directly with the accelerating voltage (Figure 5.7) and inversely with the average atomic number of the sample (Figure 5.8).

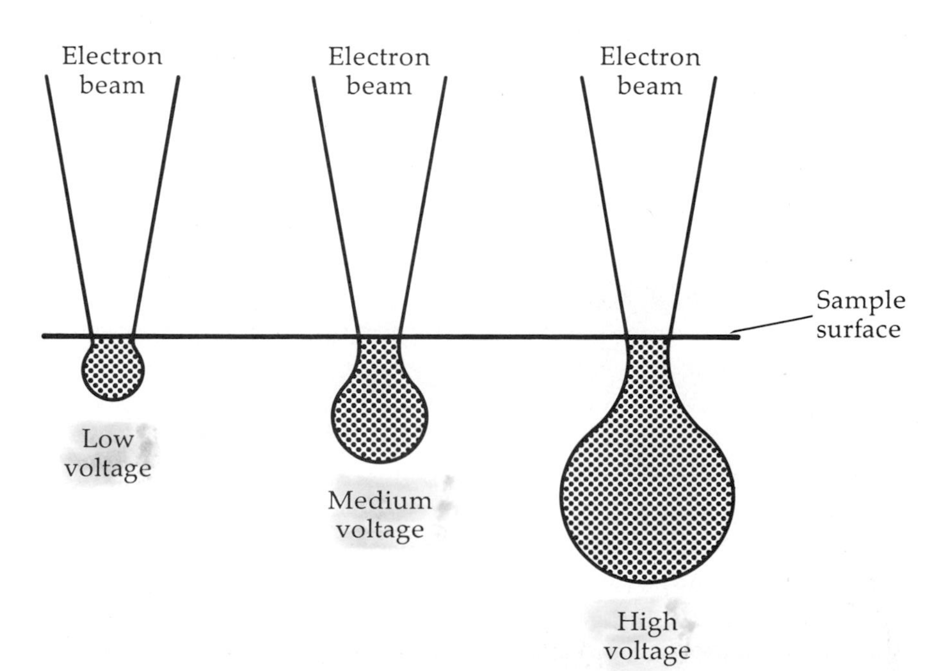

Figure 5.7 Interaction-volume variation with accelerating voltage. The accelerating voltage affects the shape, depth, and volume of the region of sample and electron-beam interaction. The higher the voltage, the larger and deeper the interaction volume.

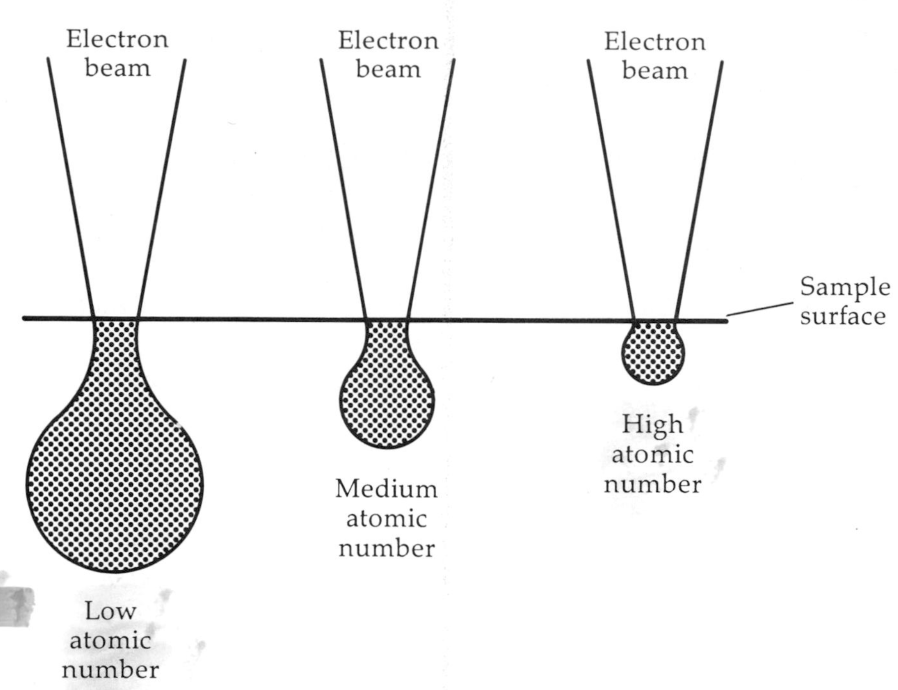

Figure 5.8 Interaction-volume variation with average atomic number of sample. The average atomic number of the sample affects the shape, depth, and volume of the region of sample and electron-beam interaction. The lower the average atomic number in the sample, the larger the interactive volume.

Interactions May Be Elastic or Inelastic

The interactions between incident electrons and the atoms of the sample may be elastic or inelastic (see Figure 4.9). Elastic interactions occur between incident electrons and the nucleus of atoms of the sample and are characterized by a large-angle deflection of the incident electron, as well as little energy loss by the incident electron. Inelastic interactions occur between the incident electrons and the orbital shell electrons of the atoms of the sample and are characterized by a small-angle deflection of the incident electron, as well as much energy loss by the incident electron.

Secondary Electrons Produce the Standard SEM Image

An inelastic interaction of major importance to SEM is the one that produces secondary electrons. Secondary electrons are produced by interactions between incident electrons and weakly bound conduction-band electrons in the atoms of the sample. The average energy of secondary electrons is about 3 eV to 5 eV. Because of their low energy, secondary electrons are attracted easily by the positive 300-V charge on the Faraday cage, and a high proportion of all secondary electrons emitted from the sample are detected, even though their initial direction may have been away from the detector. The standard image in the SEM is composed mainly of secondary electrons.

Secondary electrons are produced from the entire area of specimen–beam interaction. Because of their low energy, however, they are strongly absorbed by the sample. Only the secondary electrons produced near the surface of the specimen can escape because of the decreased path length to the surface (Figure 5.9). In general, only about one percent of all secondary electrons produced actually escape and contribute to image formation. The maximum escape depth has been calculated as 5 nm in metals and 50 nm in insulators. Most biological samples are coated with 10 nm to 20 nm of metal; therefore, the 5-nm escape depth is probably an appropriate value for the escape range of secondary electrons from such a sample.

The absorption and escape of secondary electrons are the major factors that contribute to their ability to produce a predominantly topographical image in the SEM. Small projections on the sample surface have areas of shorter path length for the escape of secondary electrons than do flat areas (Figure 5.10). Such areas appear bright on the image. Because the secondary electrons escape from a small volume of the total specimen–beam interaction volume, the secondary-electron image provides the image of highest resolution.

The Everhart-Thornley detector collects secondary electrons that have originated from four different processes (Figure 5.11). Some secondary electrons originate directly from the region of specimen–beam interaction in the sample. These electrons have the highest resolution capability because they originate from the smallest volume in the sample. A second type of secondary electron originates from the interaction of backscattered electrons with the specimen atoms as the backscattered electrons exit the sample. This type of secondary-electron signal has lower resolution, typically about that of the

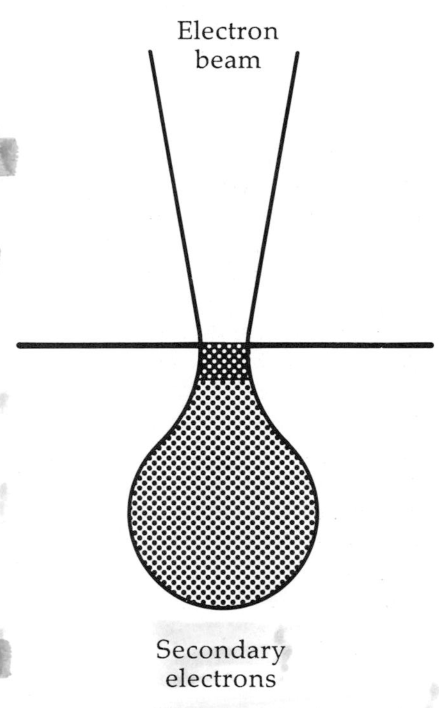

Figure 5.9 Escape depth of secondary electrons. Secondary electrons are produced from the entire region of sample–beam interaction. Only those produced in the dark-shaded region, however, are able to escape from the sample and contribute to image formation. The secondary electrons produced from deeper in the sample are absorbed by the sample and are not able to escape.

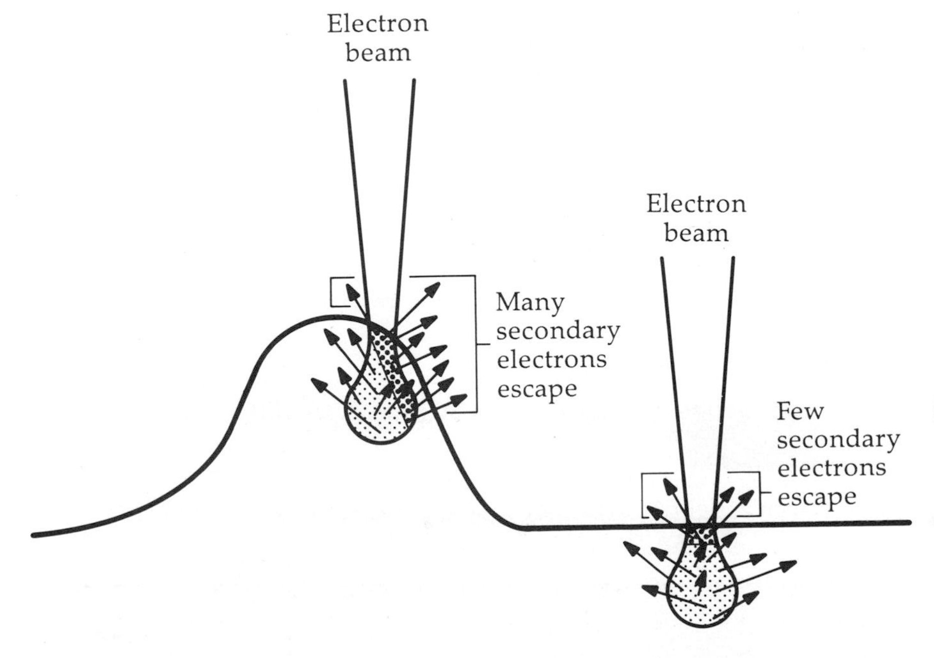

Figure 5.10 Secondary electron escape, edge effects, and image production. Secondary electrons are produced from the entire region of sample–beam interaction. Only those produced in the dark-shaded region, however, are able to escape from the sample and contribute to image formation. More secondary electrons escape from small projections than from flat surfaces because of decreased path length, a phenomenon called edge effect. The differences in the resulting voltage levels in the detector produce the image.

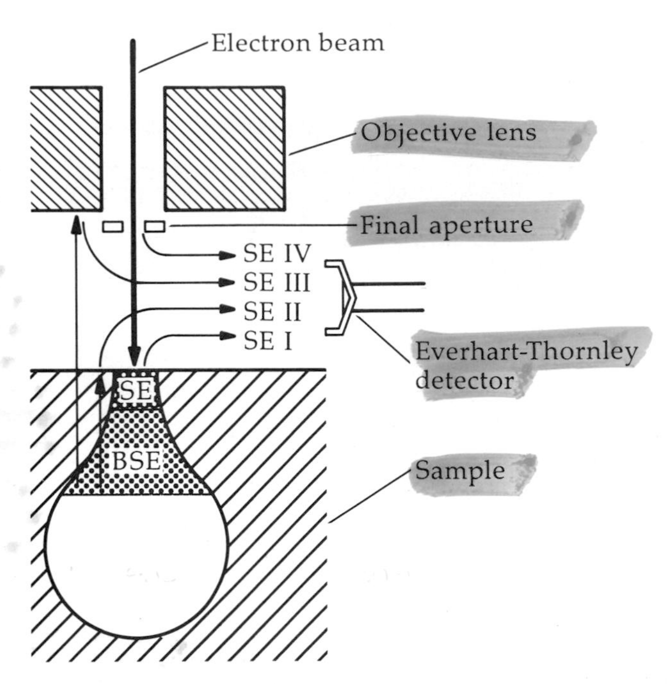

Figure 5.11 Origins of secondary electrons in an SEM. The secondary electrons (SE) that reach the Everhart-Thornley detector may originate from four different sources. Type-I secondary electrons originate directly from the area of sample–beam interaction. Type-II secondary electrons are produced by backscattered electrons (BSE) as they exit the sample. Type-III secondary electrons are produced by interactions between backscattered electrons derived from the sample and various parts of the sample chamber. Type-IV secondary electrons are produced by interactions between the electron beam and the final aperture.

backscattered electrons. Some of the backscattered electrons exit the sample without producing secondary electrons and strike the objective-lens pole piece or other parts in the specimen chamber, where they then produce secondary electrons. This third type of secondary electron has the lower resolution capability of the backscattered electrons. The fourth type of secondary electron is produced by the primary beam of electrons as they pass through the final aperture. These secondary electrons contribute an overall background noise to the image and limit resolution. Detection of only the first type of secondary electron would give the best resolution but unfortunately is not possible. The Everhart-Thornley detector is not capable of distinguishing among the four types of secondary electrons. In addition, as mentioned previously in this chapter, some backscattered electrons are also collected. Therefore, the secondary-electron image produced by the Everhart-Thornley detector is a complex mixture of electrons of different origins.

Backscattered Electrons Show Differences in Atomic Number

An elastic interaction of major importance in SEM is that which produces the backscattered electrons, i.e., beam electrons that have scattered backward. Typically 10% to 50% of the beam electrons undergo multiple large-angle elastic interactions, and they are scattered out of the sample from the same side that they entered. Backscattered electrons are high in energy, with an average of 60% to 80% of the initial energy of the electron beam. Because of their high energy, backscattered electrons are not deflected by the positive 300-V charge on the Faraday cage, and only a small percentage of the total is detected along with the secondary electrons. Efficient detection of backscattered electrons requires a special detector. At present two types exist, a semiconductor type and the Robinson type, which consists of a very large scintillator mounted directly over the sample.

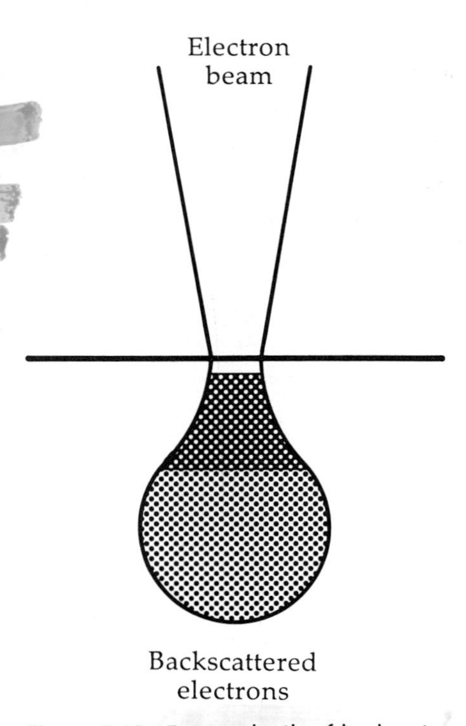

Figure 5.12 Escape depth of backscattered electrons. Backscattered electrons are produced from the entire region of sample–beam interaction except for a small area at the sample surface. Only those produced in the dark-shaded region, however, are able to escape from the sample and contribute to image formation. The incident electrons must penetrate several layers of atoms before there is sufficient probability of the multiple elastic interactions needed to scatter the electrons backward. Because backscattered electrons are very high in energy, they can escape from far deeper in the sample than can secondary electrons.

Backscattered electrons are produced from nearly the entire area of specimen–beam interaction, although few are produced from the very top layer of interaction. Scattering probability states that only one beam electron in several hundred will undergo a scattering event that diverts its path by 90° or more. A beam electron normally must be diverted more than 90° to exit the sample, which requires many large-angle scattering events. The probability of multiple, large-angle scattering events increases with sample depth. Therefore, few backscattered electrons are produced from the specimen surface.

Backscattered electrons are not strongly absorbed by the sample because of their high energy, and a high percentage do escape the sample (Figure 5.12). The maximum escape depth (and width) varies inversely with the average atomic number of the sample; the range is from a fraction of a micrometer to several micrometers. Because of the large width of escape, the resolution of an image produced using backscattered electrons is less than that of an image produced using secondary electrons. The guaranteed resolution for an SEM using backscattered electrons might be 15 nm, whereas with the secondary-electron image resolution might be 4 nm. The actual resolution of a backscattered electron image is usually considerably less than the guaranteed resolution, which is measured with an ideal sample, such as gold particles evaporated on a carbon substrate.

The large escape depth of backscattered electrons produces an image less sensitive to differences in surface topography than is the secondary-electron image, which is dependent on major absorption differences from the surface. The backscattered electron image is a depth image, and the secondary-electron image is a surface image (Figure 5.13).

The production of backscattered electrons is strongly dependent on the average atomic number of the sample. A sample with an average atomic number of 7 (typical for many biological samples) scatters about 5% of the beam electrons backward. A sample with an average atomic number of 47 (silver) scatters about 40% of the beam electrons backward. Backscattered electrons are therefore very useful in detecting the presence of differences in average atomic number of a sample. For

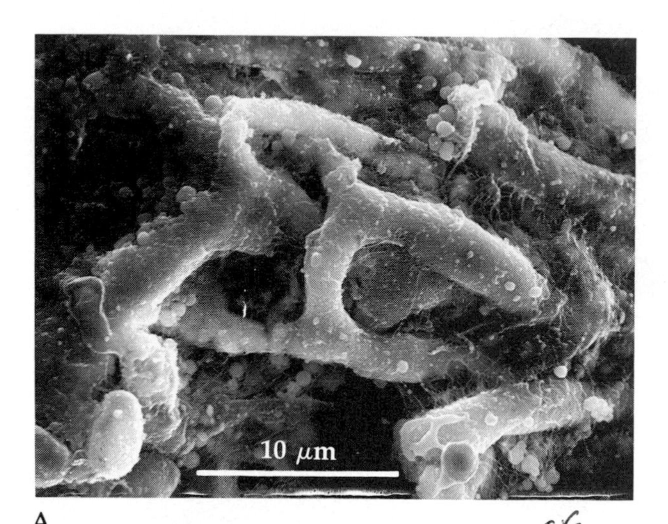

A

B

Figure 5.13 Comparison of secondary-electron imaging with backscattered electron imaging on a stained biological sample. (A) Secondary-electron image of fungal hyphae stained with a histochemical stain that deposits silver to show the location of polysaccharides. Only surface topography is visible. (B) The backscattered electron image clearly shows the location of the stain.

instance, biological tissue may be stained with heavy metals and the backscattered electron image used to reveal the location of the metal on (or in) the sample.

Flat samples often show atomic number contrast with the secondary-electron image (Figure 5.14). This effect is due to the low contrast produced by the Type-I secondary electrons (see Figure 5.11); i.e., the sample is flat, so there are few difference in absorption and thus little contrast. In such a sample, atomic number differences are apparent in the secondary-electron image because much of the contrast is derived from Type-II secondary electrons, which originate with backscattered electrons.

Several Other Products of Inelastic Scattering Are Useful in Image Formation

If an incident electron displaces an inner-shell electron, the resulting imbalance may be corrected by the production of an X-ray or by the production of a low-energy electron called an Auger electron. Measurement of the energy or the wavelength of the resulting X-ray can be used to determine the elemental composition of the sample of atomic numbers in the range of 8 to 99 (see Chapter 8) and has considerable application in biological and materials sciences. X-ray analysis requires the purchase of a separate detector and X-ray analyzer, of which several different types are available.

Measurement of the energy level of Auger electrons may also be used to determine the composition of the top 1 nm of the sample and is most useful in the range of atomic numbers from 2 to 10. Analysis of Auger electrons has little use in biology because of the requirement that

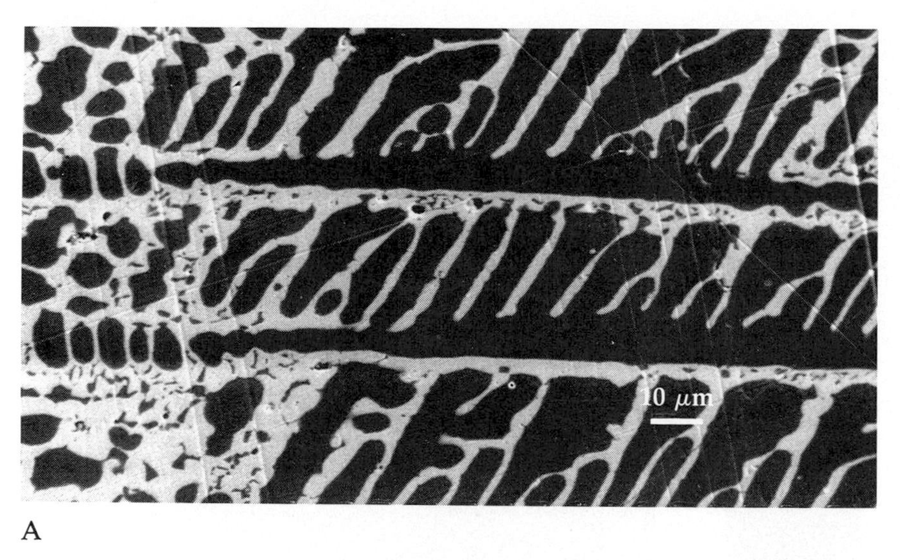

A

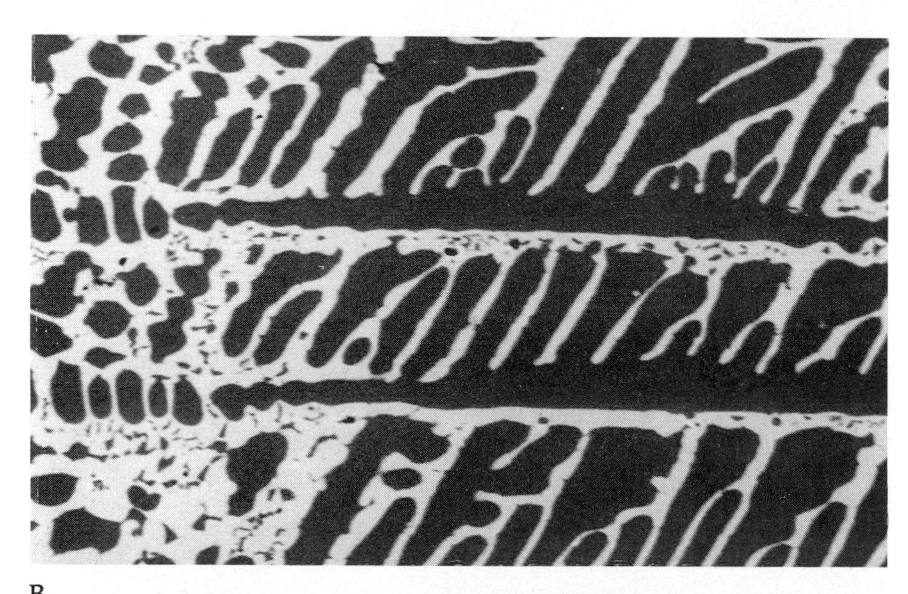

B

C

Figure 5.14 Comparison of secondary electron imaging with backscattered electron imaging on a polished and a fractured metal sample. (A) Secondary-electron image of a polished cross section of an ancient Greek coin from 300 B.C. The secondary-electron image shows atomic number differences on flat, smooth samples. In this instance, the two-constituent alloy composed of copper and silver regions is clearly revealed along with small surface scratches. (B) The backscattered electron image of the sample produces slightly more contrast than the secondary-electron image and does not show the surface scratches. (C) A secondary-electron image of a fractured cross section of the same sample shows surface topography only and does not show atomic number differences.

A

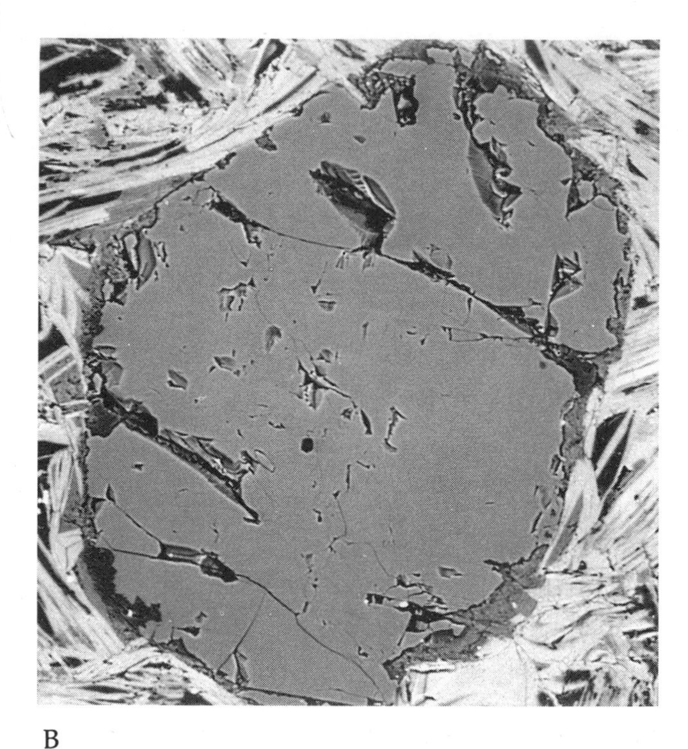

B

Figure 5.15 Cathodoluminescence. (A) A cathodoluminescence image of a polished geological sample with a central corrundum grain showing the presence of various light-emitting areas within the grain. (B) A backscattered electron image of the same sample. Photos courtesy of C. H. Nielson and Japan Electron Optics Laboratory (JEOL), USA, Inc., Peabody, MA.

the sample be inherently conductive, but it has considerable value in metallurgy, the materials sciences, and chemistry. Detection of Auger electrons requires the purchase of a special type of SEM equipped with a high-quality vacuum system and a special detector. Such a system usually is called an Auger microprobe.

A third product of inelastic scattering is the emission of light from certain solids, a process called cathodoluminescence. Conductors, semiconductors, and insulators are all capable of emitting light. The production of light is a result of electron transitions either across certain forbidden band gaps or across certain band structures called quantum wells. Cathodoluminescence has considerable application in chemistry and mineralogy (Figure 5.15) and limited application in biology. The biological applications include detection of herbicides, of calcified regions in aortae, and of histochemical stains. Cathodoluminescence detection in SEM requires the purchase of an accessory detector, of which several types are available.

Diffraction Techniques May Be Useful in Imaging

Three imaging methods rely on the principle of diffraction: selected-area electron-channeling patterns (SACP), electron backscattered patterns (EBSP), and micro-Kossel X-ray diffraction patterns (MKXDP). All of these methods provide information on crystal orientation and

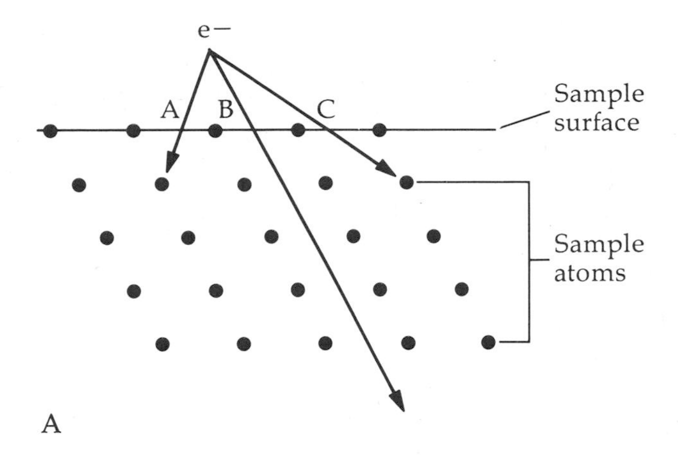

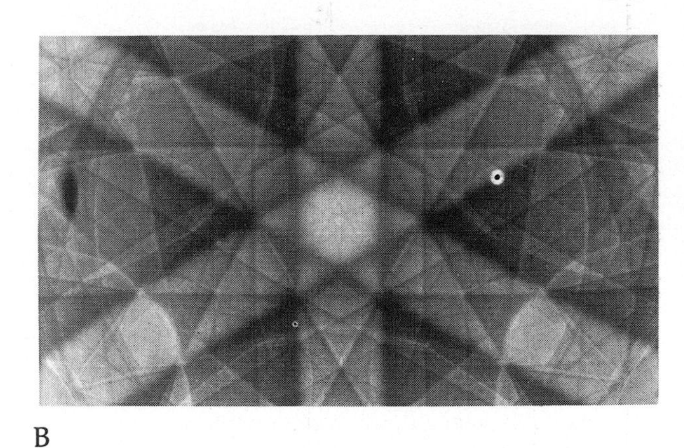

Figure 5.16 Electron channeling. (A) At certain angles of entrance, the electron beam channels deep into the sample. (B) Electron-channeling image of a silicon crystal 10 mm in diameter. Photo courtesy of V. E. Robertson and Japan Electron Optics Laboratory (JEOL), USA, Inc., Peabody, MA.

perfection, as well as lattice parameter values; thus, these techniques are of special interest to chemists, physicists, and metallurgists.

Selected-area electron-channeling patterns rely on the principle of electron channeling. When the electron beam enters a crystalline material from widely divergent angles (i.e., in a sample imaged at very low magnification), for most angles of entrance the electron beam interacts with atoms close to the surface, and a large number of incident electrons are scattered backward. Because these backscattered electrons are produced near the surface, nearly all are able to escape. If, however, the electron beam enters the sample at an angle parallel to a lattice plane, the electrons may channel deep into the lattice structure before being scattered backward (Figure 5.16A). Far fewer backscattered electrons are able to escape from deep in the sample because of absorption by the sample. Thus, a difference in contrast is produced (Figure 5.16B) whenever the beam enters at certain angles that correspond to crystal planes. Electron-channeling contrast is useful only at very low magnification when imaging large samples (10-mm square) because at higher magnifications the angular deviation of the electron beam becomes too small to locate the channels. Electron-channeling contrast can be used, however, to show the presence of grains, twins, and other crystallographic features. The various grains of a metal have the crystal planes oriented in different directions, and each grain produces a difference in electron-channeling contrast.

Selected-area electron-channeling patterns (SACP) are a much more useful form of electron-channeling contrast. With SACP, the SEM is modified so that the electron beam can be rocked back and forth through a cone of incident angles at high magnification; many modern SEMs now have this feature. SACP is, thus, capable of providing crystallographic information from much smaller areas than those for which conventional electron channeling can provide information and therefore has more potential uses.

The technique of producing electron backscattered patterns (EBSP) places a stationary electron beam on a highly tilted sample. A

Figure 5.17 Electron backscattered patterns. EBSP image of an iron/aluminum body-centered cubic derivative-ordered alloy. Photo courtesy of J. J. Stout and M. J. Crimp.

sheet of electron-sensitive film (or other electron recording device) is then placed in close proximity to the sample and parallel to the electron beam. The inelastically scattered electrons produced within the sample strike the crystallographic planes and become diffracted outward and strike the film (Figure 5.17).

Micro-Kossel X-ray diffraction patterns (MKXDP) are produced in a manner similar to EBSP, except that a window of material transparent to X rays, such as a thin film of beryllium, is placed between the sample and the film. The window prevents the passage of electrons, but allows the passage of X rays. The X rays produced within the sample from bombardment by the electron beam then become diffracted outward and strike the film.

Each of the techniques just described has certain advantages. EBSP has a spatial resolution of 20 nm and an angular resolution of 1° to 2°. SACP has a spatial resolution of 1 μm and an angular resolution of 0.1° to 0.2°. MKXDP has a spatial resolution of 100 nm and an angular resolution of 0.01°. Both EBSP and SACP require a carefully polished sample; MKXDP can be used with rough samples.

Other Methods of Image Production

Four other methods exist for image production in an SEM: sample current, voltage contrast, electron beam–induced current (EBIC), and magnetic contrast.

Sample-current or absorbed-current imaging relies on the presence of internal variations in current flow while a sample is being scanned (Figure 5.18). This image is relatively independent of surface

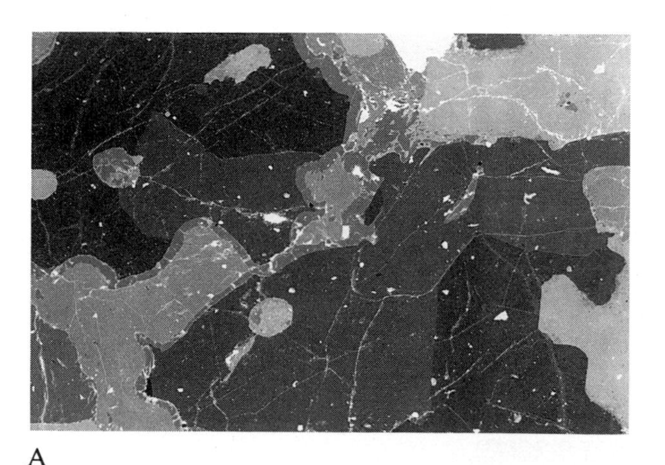

A

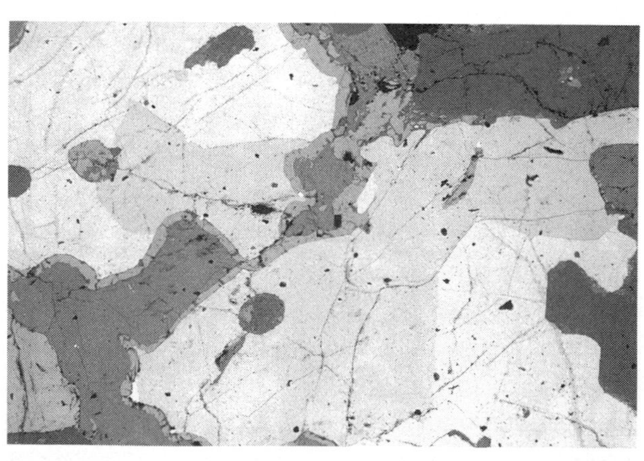

B

Figure 5.18 Sample-current imaging. (A) A sample-current image of polished felsitic rock. (B) A backscattered electron image of the same sample. A sample-current image normally is inverted contrast of the backscattered electron image. Photos courtesy of V. E. Robertson and Japan Electron Optics Laboratory (JEOL), USA, Inc., Peabody, MA.

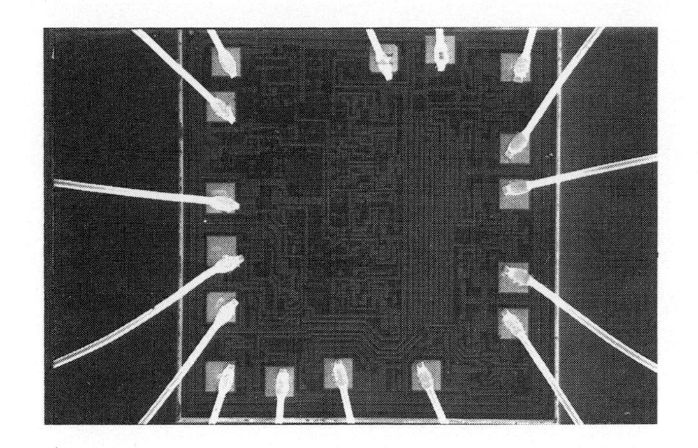
A

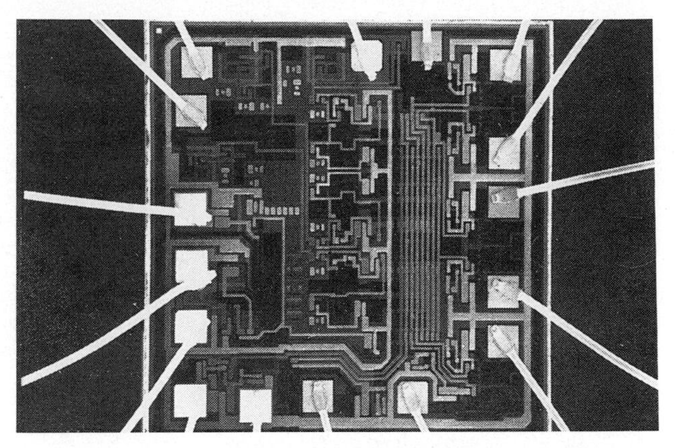
B

Figure 5.19 Voltage-contrast imaging. (A) A secondary-electron image of an LS 138 multiplexer integrated circuit chip with no applied voltage on the device. The circuit pathways are a uniform shade of gray. (B) The device with applied voltage. The circuit pathways display different shades of intensity that are related to the applied voltage levels. Photos courtesy of R. S. Cornell and Japan Electron Optics Laboratory (JEOL), USA, Inc., Peabody, MA.

variations; mainly internal variations are visualized. Sample current has some use in the fields of materials science and mineralogy and few applications in biology. Sample-current detection requires the purchase of a special amplifier for an SEM.

Voltage-contrast imaging is a result of the effect that a small voltage placed on a small area of a sample has on the brightness of that area. If a positive voltage is placed on a certain portion of a sample, fewer secondary electrons are able to escape, and that portion of the sample thus appears darker than most surrounding areas. Conversely, a negative voltage causes a portion of a sample to appear brighter. Voltage-contrast imaging is especially useful in the development of semiconductors and integrated circuits because it is possible to observe the various circuit pathways while the device is in operation (Figure 5.19).

Electron beam–induced current (EBIC) can be obtained when an electron beam is focused on a semiconductor p-n junction. The incident electrons create excess electron-hole pairs that produce a current in that junction. The current can be used for the video signal in the SEM, and thus, can be used to generate an image (Figure 5.20). EBIC is used mainly in the imaging and studying of semiconductors.

Two types of magnetic contrast can be used in imaging. With one type, magnetic fields above the sample deflect the secondary electrons emitted from the sample and produce changes in contrast. This mechanism has been used to view the magnetic domains on recording tape. The other type of magnetic contrast occurs when internal magnetic fields affect the scattering of the incident electrons. The backscattered electrons in this instance can be used to show differences in magnetic domains. Both methods are useful in studying all types of magnetic memory devices.

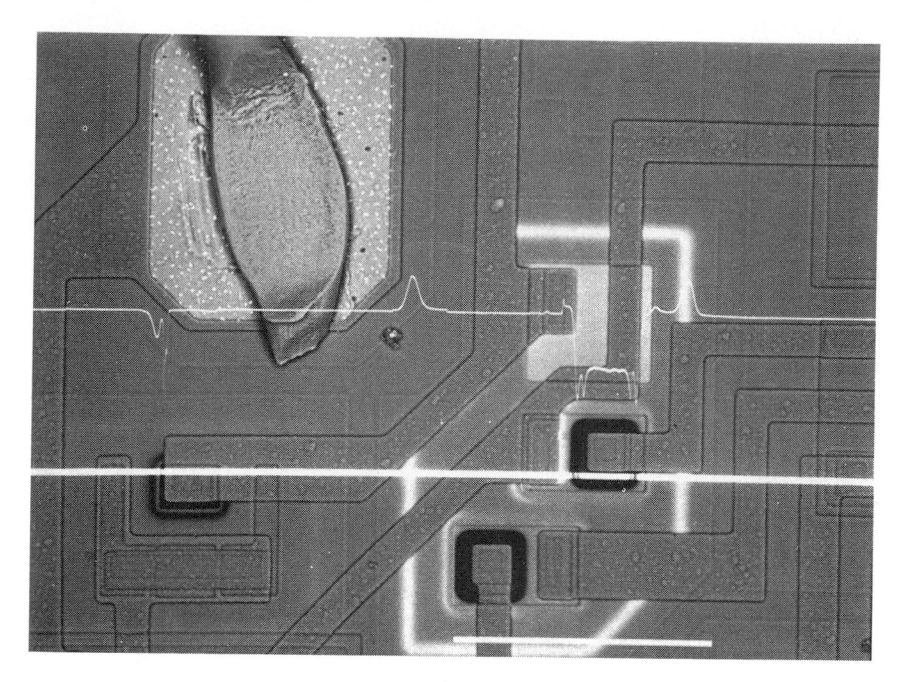

Figure 5.20 Electron beam–induced current. An electron beam–induced current (EBIC) image of a type-741 operational amplifier integrated circuit chip. The active semiconductor region appears as bright bands. The thin horizontal line indicates the EBIC signal level on a linescan; the thick horizontal line indicates the location of the linescan itself. Photos courtesy of R. S. Cornell and Japan Electron Optics Laboratory (JEOL), USA, Inc., Peabody, MA.

MACHINE VARIABLES

Several adjustments on the SEM, usually called machine variables, have a major impact on the results. Proper use of these variables often determines whether the final image is adequate in conveying the information desired. Understanding how changing the variables produces desirable and undesirable effects is also important. As with the TEM, a balance of trade-offs is often required.

Condenser-Lens Adjustments Affect Resolution

The objective lens (or focusing lens) produces a crossover or focus on the sample. Increasing the current in the lens moves the focus up on the Z axis, and decreasing the current moves the focus down on the Z axis. The crossover does not produce an infinitely small spot size but instead produces a narrowed spot, the diameter of which is controlled by changing current in the condenser lens. Increasing the current in the condenser lens reduces the focal length of the crossover produced between the condenser lens and the objective lens, and results in a spot of smaller diameter on the sample. Reducing the focal length of the condenser lens, however, causes the angle of divergence of the electron beam (aperture angle) to become wider, and less of the beam is able to

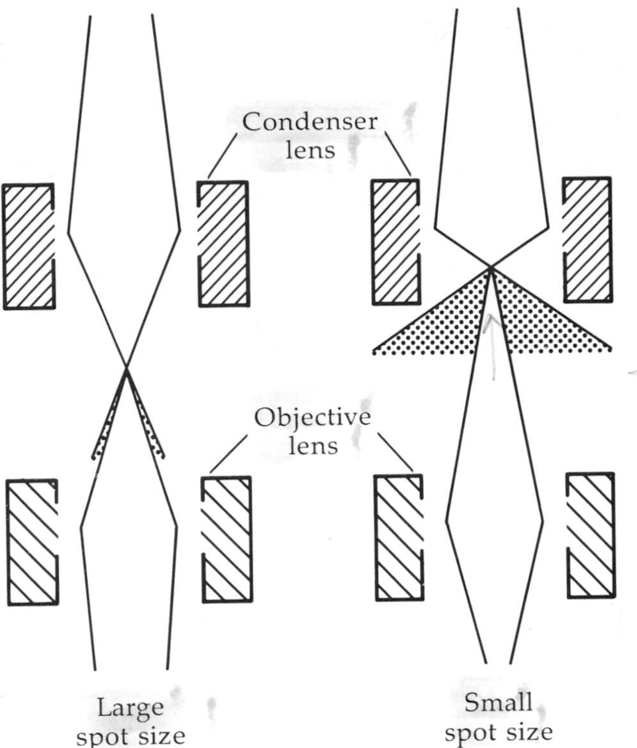

Figure 5.21 Spot size and beam current. When the condenser lens current is increased to produce a small crossover and thus a small spot size, the objective lens intercepts less of the electron beam (the stippled area), resulting in lower beam current.

enter the aperture of the objective lens, resulting in decreased beam current (Figure 5.21).

In general, the smaller the spot size, the better the resolution will be (Figure 5.22). It is not possible to resolve any two structures whose

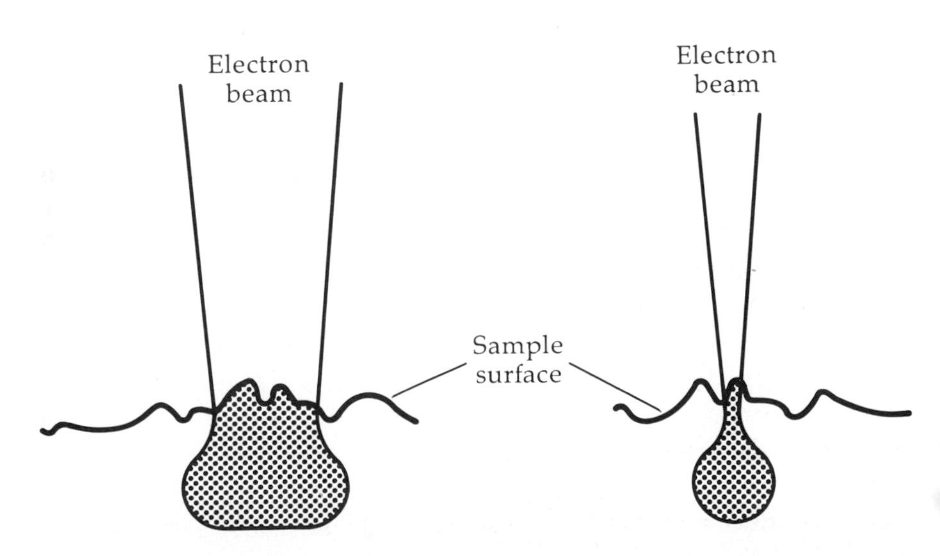

Figure 5.22 Spot size and resolution. The diameter of the beam of electrons as it scans the sample, called the spot size, is directly related to the resolution. In general, the smaller the spot, the greater the resolution.

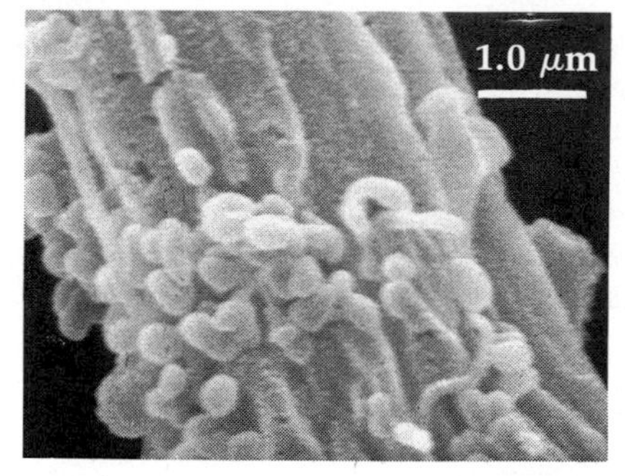

A

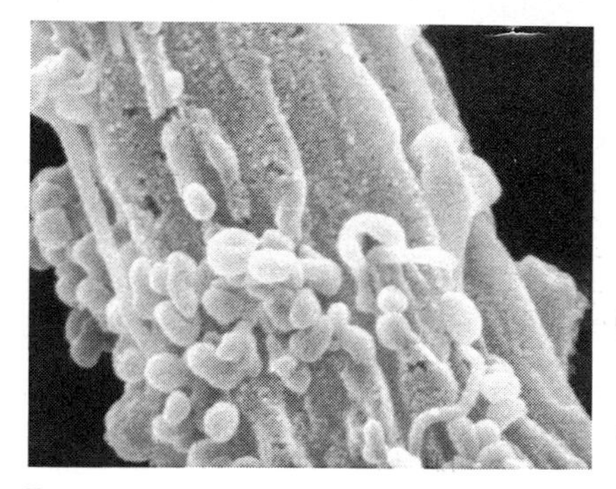

B

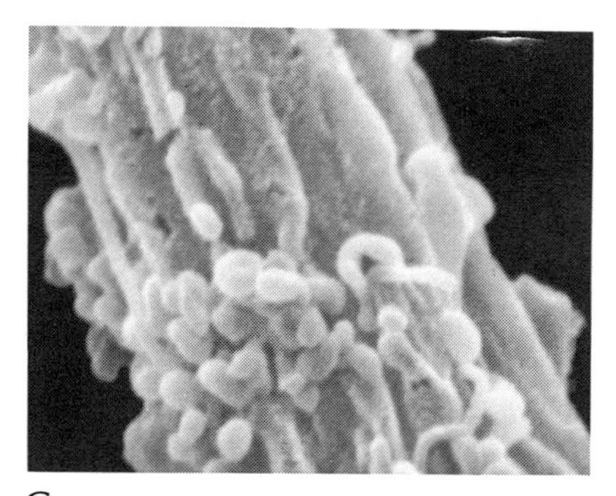

C

Figure 5.23 Spot size, resolution, and image quality. (A) A macrophage (a type of blood cell) photographed using the smallest spot obtainable on a particular SEM. (B) The same cell photographed using a slightly larger spot size than the previous example. Note the improvement in resolution. (C) The same cell photographed with a large spot size. Note the poor resolution.

spacing is less than the diameter of the beam of electrons. As the diameter of the spot is reduced, however, the beam current is reduced; therefore, the quantity of secondary and backscattered electrons decreases, resulting in a weaker signal from the detector. To compensate for this problem, the gain of the photomultiplier must be increased to give sufficient contrast. This adjustment results in greater electronic noise from the photomultiplier, which appears as a graininess in images and photographs. Eventually, if the grain becomes excessive, it limits the resolution (Figure 5.23). A balance between spot size and adequate signal level must be achieved. This balance varies from sample to sample and with the other machine variables.

Working Distance and Aperture Size Affect Depth of Field and Resolution

The beam of electrons is cone-shaped when it strikes the sample. The angle of the cone (twice the aperture angle α, see Figure 2.9B), and whether it is a wide cone or a narrow cone, is determined by the diameter of the final aperture and the working distance. Most SEMs allow the selection of three different sizes of final apertures. A final aperture with a small diameter produces a narrow cone of electrons; one with a large diameter produces a wide cone (Figure 5.24). Working

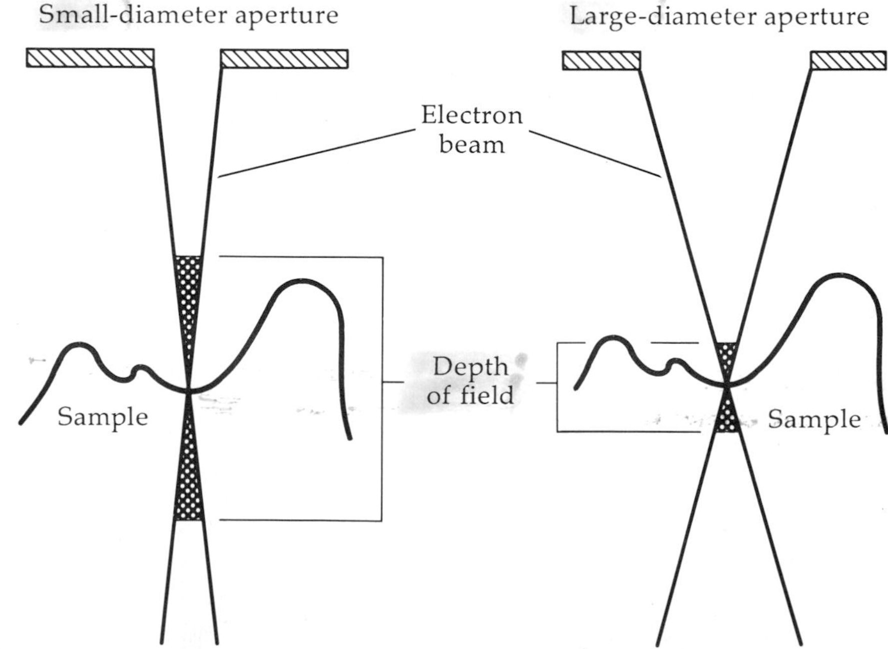

Figure 5.24 Aperture size and depth of field. The beam of electrons has a cone shape when focused on the sample. A maximum spread of the cone may be defined as the region in acceptably sharp focus; this is called depth of field. The depth of field is greater with a final aperture of smaller diameter.

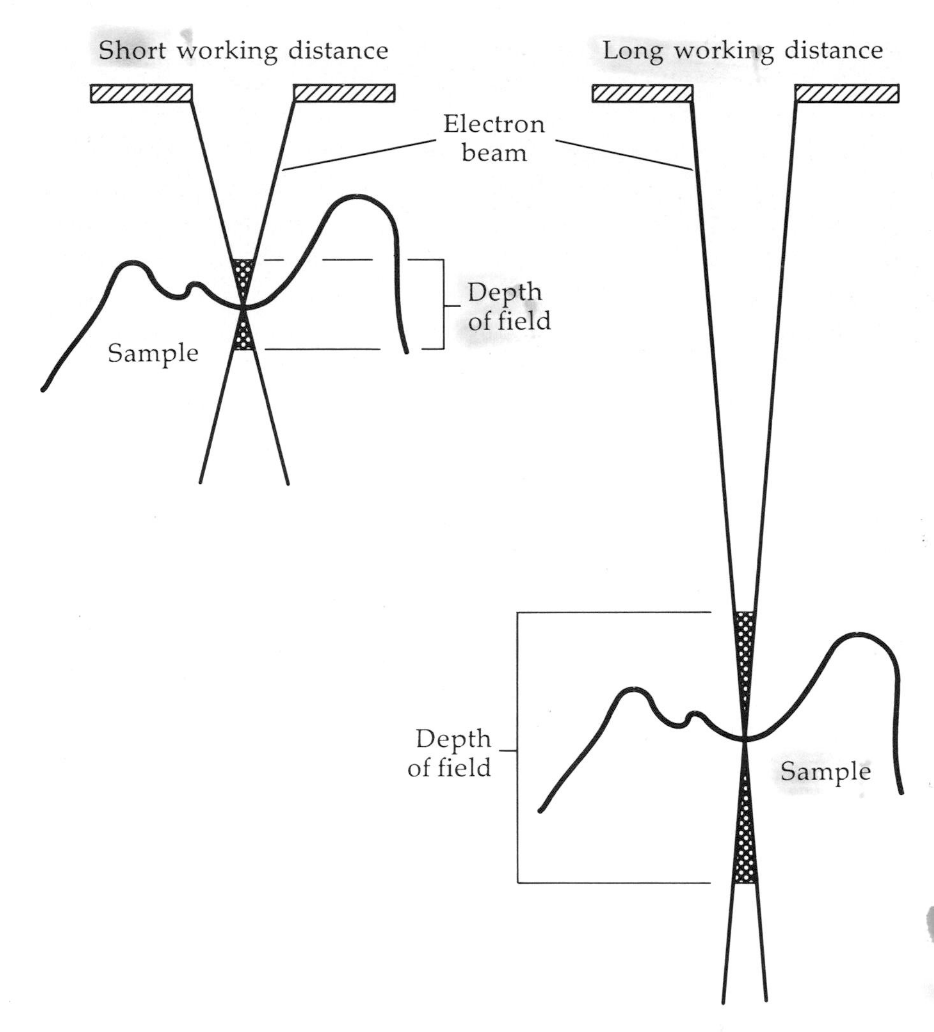

Figure 5.25 Working distance and depth of field. The depth of field is greater at long working distances than at short working distances.

distance is the distance between the objective lens and the sample. Most SEMs have a working distance that varies from about 10 mm to 40 mm or 50 mm. A large working distance results in a narrow cone of electrons, a small one in a wide cone (Figure 5.25).

The angle of the cone is the most important factor in determining the depth of field at any given magnification. Depth of field is the amount of sample that is in acceptably sharp focus. The effect of the angle of the cone on depth of field is due to the fact that the focused beam of electrons normally scans not a flat sample, but a sample with an irregular surface. The beam, however, is brought to a fixed focus, or crossover. The diameter of the beam that strikes each point of the sample varies with the topography of the sample (see Figures 5.24 and 5.25). If a sample with large topographical variation is scanned, the areas greatly above and below the crossover (focal plane) will be imaged with a beam of greatly increased diameter, resulting in reduced resolution. At a certain point, the areas of reduced resolution will appear out of focus. A certain maximum spread of the beam can be defined that will result

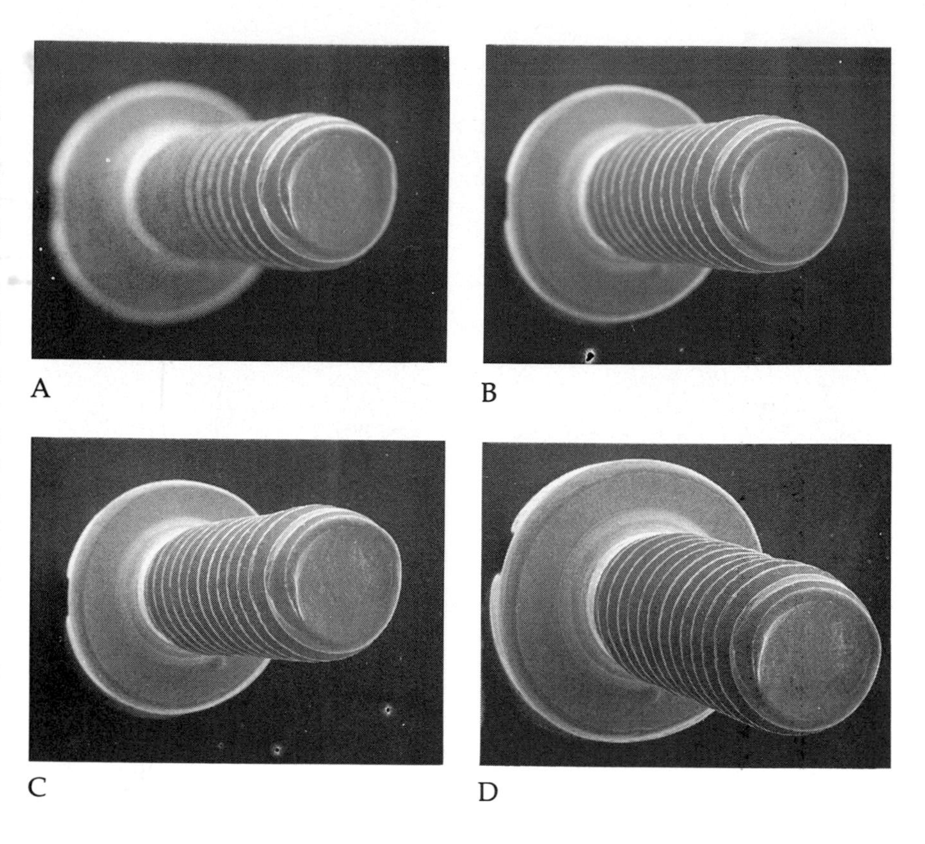

Figure 5.26 Effect of aperture size and working distance on depth of field. A machine screw photographed with different parameters. (A) 15-mm working distance, 600-μm aperture. (B) 15-mm working distance, 200-μm aperture. (C) 15-mm working distance, 100-μm aperture. The depth of field is greater with smaller aperture size. (D) 39-mm working distance, 100-μm aperture. The depth of field increases with smaller aperture size and with longer working distances. Combined, a longer working distance and an aperture of smaller aperture give the greatest depth of field. The tip of the screw appears to be larger in relation to the size of the head at a 15-mm working distance than it does at a 39-mm working distance. This is an artifact similar to the distortion observed in a wide-angle camera lens when a close object and an object far away are in the same view. In addition, a slight rotation of the image is observed between the two working distances.

in adequate resolution; therefore, calculations and comparisons may be made between working distances and objective-lens apertures of different diameters.

Using a final aperture with a smaller diameter produces more depth of field; halving the diameter doubles the depth of field, and doubling it halves the depth of field (Figure 5.26). A final aperture of smaller diameter also reduces the intensity of the beam, resulting in lower signal levels; in some instances this may be a major factor limiting the quality of the results.

A longer working distance decreases the angle of the beam cone and results in greater depth of field. Doubling the working distance doubles the depth of field; halving it halves the depth of field. Unlike using the objective-lens aperture to increase depth of field, increasing the working distance only slightly reduces signal levels. Varying both aperture size and working distance can increase depth of field greatly. Changing the angle of the cone by varying either the working distance or the diameter of the final aperture affects the resolution that can be obtained from the SEM. Lenses have less spherical aberration at shorter working distances, resulting in a spot size of smaller diameter and thus better resolution. Most SEMs give better resolution with an intermediate-size final aperture. In theory, the smallest aperture should produce the best resolution because of the reduction of lens aberrations. The

Figure 5.27 Effect of aperture size and working distance on resolution. Magnesium hydroxide crystals with an organic coating photographed with different combinations of variables. (A) 15-mm working distance, 100-μm aperture. (B) 15-mm working distance, 200-μm aperture. (C) 15-mm working distance, 600-μm aperture. (D) 39-mm working distance, 200-m aperture. Resolution is better at shorter working distances primarily because of decreased lens spherical aberrations. Resolution is best with the 100-μm diameter aperture at the 15-mm working distance. Better images are often obtained with an intermediate-size aperture at the 15-mm working distance because of the reduction in beam intensity with the smaller aperture that results in increased electronic noise in the images.

smallest aperture usually produces so much reduction in beam current and therefore secondary-electron production, however, that electronic noise becomes a limiting factor. Hence, the conditions that give the greatest depth of field (objective aperture with a small diameter and long working distance) also result in the poorest resolution (Figure 5.27). A balance must be obtained between the required depth of field and resolution.

Accelerating Voltage

The accelerating voltage in most modern SEMs varies between several hundred volts and forty thousand volts and has a tremendous effect on the images produced. In theory, a higher voltage should give better resolution because the reduction in wavelength of the beam of electrons enables a spot size of smaller diameter to be obtained. Unlike in the TEM, however, as the beam of electrons interacts with the SEM whole sample, there is a zone of interaction deep within the sample.

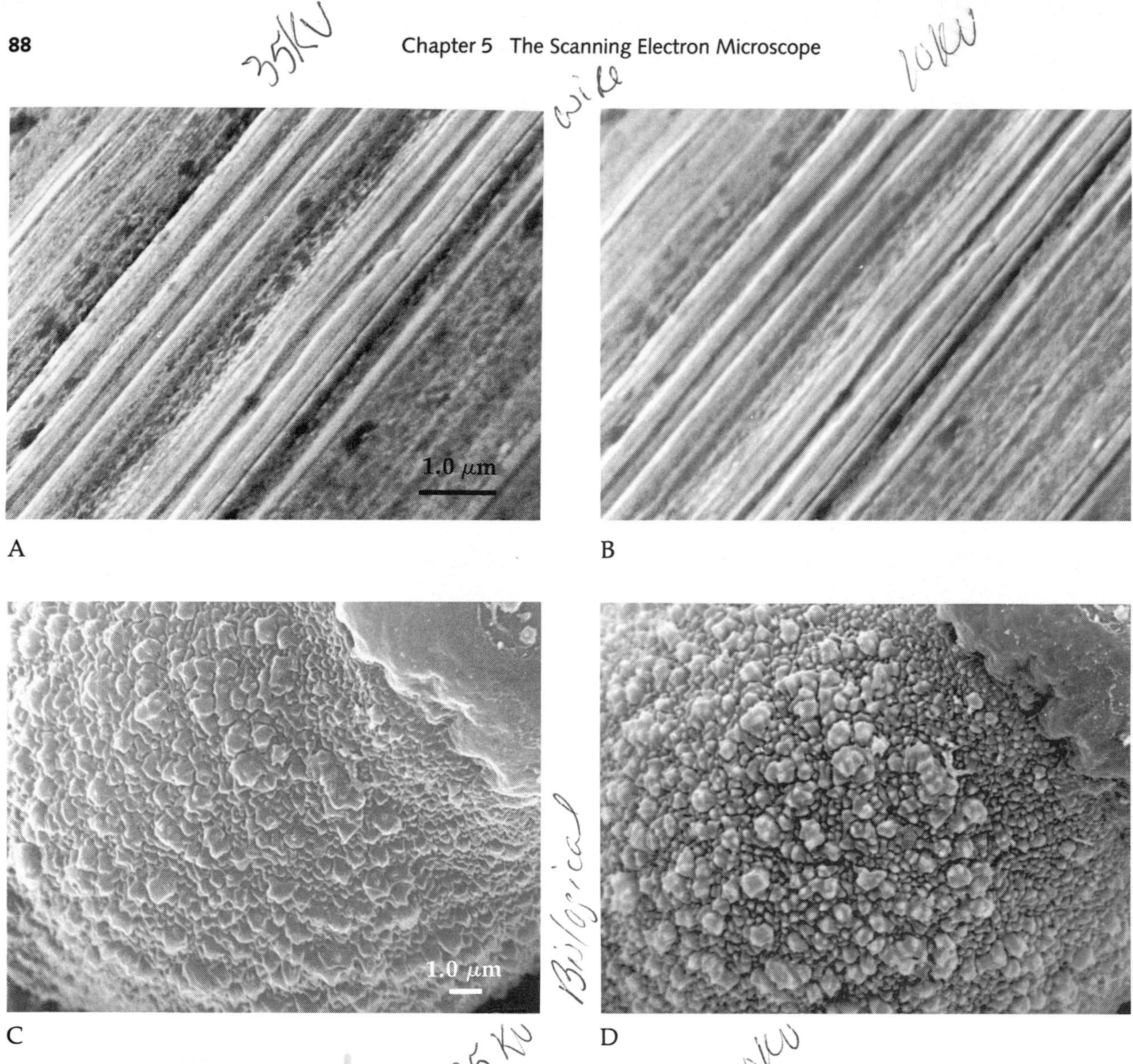

Figure 5.28 Effect of accelerating voltage on image quality. (A) Pure gold wire photographed at a 35-kV accelerating voltage. (B) The same wire photographed at 10 kV. In this instance, the best image is obtained at 35 kV. (C) Red-oak pollen photographed at 35 kV. (D) The same pollen grain photographed at 10 kV. Note the vastly improved image with much detail not evident at 35 kV.

The volume of this region increases with increasing accelerating voltages (see Figure 5.7) because of the greater energy of the beam of electrons. The increase in size of this region varies with the density of the sample and is significantly greater in samples with low density, such as those encountered in biology. The increase in volume of the region of interaction results in a decrease in resolution. In practice, a balance must be achieved in selecting the optimum accelerating voltages. In general, hard specimens like metallurgical specimens, teeth, and bones may benefit from higher voltages (Figure 5.28A and B). Most biological samples, on the other hand, produce better images at lower voltages (5 kV to 15 kV) (Figure 5.28C and D). Because of recent improvements in electron guns and in anode and lens design, many SEMs now have acceptable levels of resolution operating with accelerating voltages as low as several hundred volts. Increasing emphasis is being placed on analysis of biological and polymer samples at these very low voltages.

ULTRAHIGH-RESOLUTION SEMs

The resolution of most SEMs is about 4 nm to 6 nm. Some instruments on the market, however, are capable of much better resolution. One method of improving resolution is to use a field-emission gun. The advantages of this gun (see Chapter 2), such as its low energy spread and bright beam, enable the use of a very small spot size with reduced chromatic aberration, which results in a resolution of 1.5 nm to 2 nm. A second method of resolution is to place the specimen physically within a split objective lens of the type used in TEMs. The specifics of this arrangement are discussed in this chapter in the section on the nondedicated STEM. Resolution is between 1.5 nm and 2 nm. The best resolution can be obtained by combining the two methods just described. Machines equipped to do so are capable of obtaining resolution as low as 0.8 nm, which is very close to that of a TEM.

ENVIRONMENTAL SEMs

One disadvantage of the SEM is that it is normally not possible to examine samples that produce any significant amount of vapor when placed in a vacuum. Because of this limitation, biological samples must be dried, and many samples like grease, adhesives, liquids, foods, gels, and other semisolids cannot be examined. Some vapor-producing samples can be examined using cryogenic SEM (see Chapter 7). However, even cryogenic SEM cannot be used to observe the drying process of adhesives, the curing of cement, the melting of alloys, or the crystallization of materials.

In attempts to overcome these disadvantages, progress has been made in recent years in perfecting the environmental SEM. The environmental SEM maintains the sample chamber in a near-atmospheric environment more conducive to examination of wet samples and has a completely different environment (high vacuum) in the remainder of the column.

The entire column cannot be kept at low vacuum because the presence of large quantities of air molecules would impede the electron beam. The sample chamber, however, may be kept at a vacuum much lower than the rest of the column. The electron beam is able to traverse short distances of a low-vacuum environment with no deleterious effects. The environmental SEM is capable of maintaining a vacuum of 1.3×10^{-5} Pa (10^{-7} torr) in most of the column and a vacuum of 1.3 Pa to 2.7 × 10^3 Pa (10^{-1} torr to 20 torr) in the sample chamber by using a series of small apertures (differential apertures) in the column.

Low vacuums in the range of 1.3 Pa to 2.7 Pa (10 torr to 20 torr) allow the examination of many biological and food samples without significant drying. If the examination protocol requires observation of the drying process on certain sample types, the vacuum level may be increased. Crystallization processes can be observed easily, and a specimen heater can be used to observe the melting or solidification of metal alloys.

Although the electron beam may traverse a low-vacuum environment for a short distance, the secondary electrons cannot do so because of their low energy. In addition, the presence of a positive 12-kV charge on the scintillator would cause problems in the low-vacuum environment, thus precluding the use of a conventional Everhart-Thornley detector. One manufacturer has solved this problem by using a charged plate mounted directly above the sample. The plate carries a positive 300-V to 500-V charge. The charge tends to facilitate the ionization of molecules in the chamber. With this process, the secondary electrons released from the sample cause the ejection of electrons from the gas molecules, which then cause further ejection of electrons in a cascading manner. Water vapor injected into the sample chamber has been found to enhance this process. One major advantage of the process is that the free positive ions produced help to neutralize the buildup of a negative charge on the sample (see Chapter 7), and nonconducting samples may be examined without a metal coating. Another manufacturer has solved the detector problem by not detecting the secondary electrons and instead using a semiconductor detector to detect the backscattered electrons that are not affected by the low-vacuum environment because of their high energy level.

SCANNING TRANSMISSION ELECTRON MICROSCOPES

Advantages of STEMs

A scanning transmission electron microscope (STEM) is an electron microscope that can be used to examine thin sections and other samples in a scanning mode. Incident electrons from the beam are transmitted through the specimen in much the same manner as in the TEM. Instead of the beam of electrons being in the imaging or flood-beam mode, however, it is focused to a point source and scans the specimen in the same manner as in an SEM. Normally, the sample is scanned, and the transmitted image is produced on a CRT in the same manner as in an SEM.

Compared to the TEM, the STEM has several advantages: Sections of greater thickness may be examined, better analytical resolution can be obtained in X-ray analysis because of the small diameter of the beam, and the digitizing of information is facilitated. The ability to view thicker sections is due to the manner in which the image is created. All the energy of the beam is focused on one very small area at a given instant. Any electrons that can exit the other side activate the detector. The detector can be activated by primary electrons that are unscattered, inelastically scattered, or elastically scattered. Thus, because of the greater intensity of the beam and the ability to detect more types of electrons, thicker specimens can be viewed in the STEM than can be viewed in the TEM. The range of thickness that can be examined in an STEM varies depending on the accelerating voltage and the characteristics of the sample. In general, the thickness range is between 100 nm and 0.5 μm to 1 μm.

Often the sections do not have to be stained for the STEM because the image production process is more sensitive to variations in sample density and composition than in the TEM. In some situations, examining stained and unstained sections provides complementary information, as long as the precise details of the staining reaction are understood. The use of sections of greater thickness may give more information simply because more material is being viewed at a given time.

The ability to view thicker sections has some advantages in energy-dispersive X-ray analysis. The production of X rays is directly dependent on the number of atoms being energized by the beam. Greater section thickness causes more atoms to be energized, and therefore more X rays are produced, resulting in easier analysis. The increase in section thickness usually does not significantly reduce the analytical spatial resolution capabilities (i.e., the ability to resolve elements located in different areas).

One additional advantage of the STEM is that, because the image is a scanned image, the information may be digitized, processed, and stored in a computer. Most instruments require accessory equipment, called digital beam control, for the digitization.

Nondedicated STEM

The most common type of STEM is a standard TEM that has been fitted with a scanning attachment. This type of instrument is capable of performing TEM, STEM, and in addition will do standard SEM on small whole samples with very high resolution. This instrument has all the advantages of the STEM previously discussed, as well as the added SEM capability.

Sample size in SEM is usually about 3 mm by 8 mm, and sample movement is limited. The SEM resolution of the nondedicated STEM, in the range of 1.5 nm to 2 nm, is much better than that with conventional SEMs. The improved resolution is a result of sample placement and the detection process used for the secondary electron. The SEM sample is placed in the field of a split objective lens as is used for TEM samples. This arrangement limits objective-lens spherical aberration and increases resolution. In addition, the detection process limits the detection of two types of secondary electrons that limit resolution in a conventional SEM (see the section on secondary electrons in this chapter). Since the detector cannot be placed directly next to the sample because of the lack of room within the area of the sample, the magnetic field of the lens is used to attract the secondary electrons up through the bore of the objective lens, where they are detected by a detector placed on the top of the lens. This arrangement greatly reduces the number of Type-III and Type-IV secondary electrons (see Figure 5.11). Thus, two of the resolution-limiting signals are reduced, and the resulting resolution is better than that of a typical SEM.

The Dedicated STEM Is a Sophisticated Instrument for Examining Molecular Structure

The dedicated STEM is dedicated to the STEM mode and cannot produce a standard TEM image because it lacks post-specimen imaging lenses. Dedicated STEMs are equipped with a field-emission gun. The images from a dedicated STEM are produced in a manner considerably different from that of other STEMs. They are based on the specific type of interaction between the beam electrons and the atoms of the sample. The products of interaction (undeviated electrons, elastically scattered electrons, and inelastically scattered electrons) are detected separately, and each can be used to produce an image. Wide-angle, elastically scattered electrons are detected by the annular detector, the output of which is used to produce an image called the dark-field image. The electrons that have not been scattered through wide angles pass through the annular detector and then through an electron energy loss spectrometer.

This device is a curved magnet that separates electrons according to their velocity, which is related to energy loss. A detector placed behind the spectrometer can be adjusted to detect any range of energy loss desired. If the detector is adjusted to detect electrons with no energy loss (the type that are undeviated or transmitted as they pass through the sample), the image obtained is called the bright-field image. If the detector is adjusted to detect electrons with greater energy loss, the image is based on the detection of inelastically scattered electrons.

The high resolution (0.1 nm) and the ability to base an image on specific types of specimen–beam interactions make this instrument very useful to physicists and chemists for researching the basic structure of matter. The instrument has not been used extensively for examining biological thin sections but has been used to map the mass of proteins and nucleic acids.

SEM With Transmitted Detector

Most SEMs can be equipped with a specimen holder that holds grids and with an extra Everhart-Thornley detector located beneath the grid holder to detect the signals generated so that they are capable of STEM-type imaging. Because of the limited range of acceleration voltage of most SEMs, this type of instrument cannot be used to examine sections as thick as those that can be examined with other types of STEMs. This type of STEM is not used extensively.

SCANNING TUNNELING AND ATOMIC FORCE MICROSCOPY

The scanning tunneling microscope (STM) and the atomic force microscope (AFM) are two new types of microscopes that have great potential to complement the capabilities of the SEM. Neither the STM nor the AFM is an electron microscope; instead of using a beam of electrons to image the sample, a very fine mechanical probe scans the surface of the sample. In addition, neither instrument requires that the sample be in a vacuum. The microscope typically is no more than 15 cm to 20 cm high and can fit easily on a table.

Both the STM and the AFM typically are used to scan the surface of whole, nonsectioned samples in the same manner as that of an SEM. The advantages of the STM and AFM, however, are higher resolution and greater sensitivity to define profile differences in a sample. The resolution of the STM and AFM is in the range of 0.1 nm or less, compared to the 5-nm resolution of an average SEM. The greater sensitivity to profile differences refers to the vertical (height) variations in a sample, as compared to the lateral (horizontal) differences that normally are used to define resolution. Because of the nature of the imaging process, the STM and AFM are far more sensitive to these profile variations. In fact, some samples that show height variations barely discernable in an SEM produce an excellent image in an STM or AFM.

The scanning mechanism in both the STM and the AFM is a fine mechanical probe that scans the surface of the sample. Both instruments use a piezoelectric device for scanning. The piezoelectric effect occurs because certain crystals increase in size when a voltage is applied. By combining crystals, movement in X, Y, or Z directions is possible. In an STM, either the tip or the sample may be moved using the piezo device. In an AFM, the sample is mounted on the piezo device and moved.

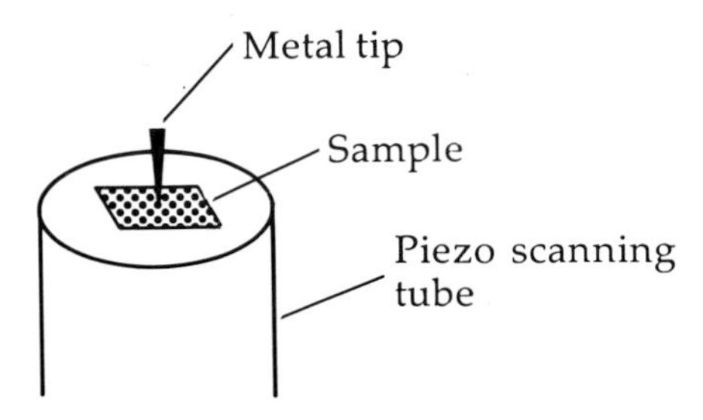

Figure 5.29 Schematic of a scanning tunneling microscope. The image produced by an STM is based on the current that flows between the metal tip and the sample.

In an STM, a metal tip is positioned very close to the surface (Figure 5.29). The tip is so close that the orbital shells of the electrons in the tip overlap with those in the sample. If a bias voltage is applied between the tip and the sample, some of the electrons are able to tunnel between the tip and the sample, and thus, cause a flow of current. The variation in current as the tip moves can then be used to produce an image on a CRT in a manner similar to that in an SEM. Alternatively, a feedback mechanism can be used to move the tip in a Z direction to keep the tunneling current constant as the tip scans the sample surface. In this instance, the variation in voltage that must be applied to the piezoelectric device to keep the current constant is used to produce the image on the CRT. Because the tunneling current is extremely sensitive to variations in distance between the tip and the sample, the image shows high contrast with very small variations in surface height. It is important that the sample be conductive for proper operation of the STM.

In an AFM, a very fine tip is mounted on a triangular piece of metal foil called the cantilever (Figure 5.30). The piezoelectric device moves the sample under the tip. The variation in attractive forces between the electrons in the orbital shells of the tip and those of the sample cause movement of the foil. A laser beam is also directed at an angle toward the surface of the foil. The reflected beam of the laser is detected by a photodiode. Movement of the foil causes variation in the current in the photodiode. This variation in current is then used to produce an image on a CRT. Alternatively, a feedback mechanism can be used to move the tip to keep the photodiode current constant, and the variation in voltage applied to the piezoelectric device can be used to produce the image. One major advantage of the AFM is that samples do not need to be conductive.

The STM and AFM have great potential in many fields of research, both biological and nonbiological (Figure 5.31). In nonbiological sam-

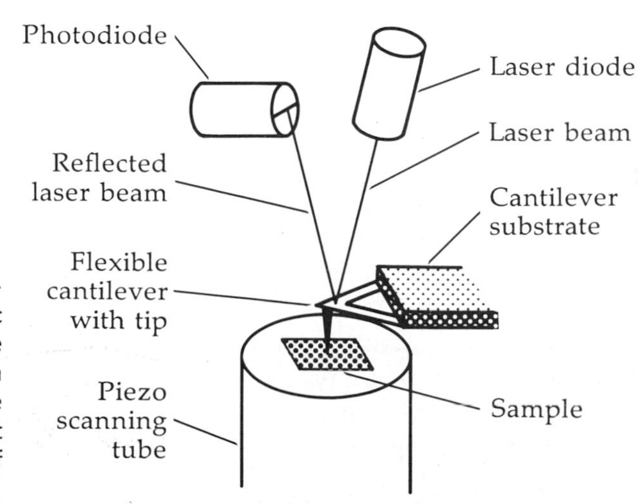

Figure 5.30 Schematic of an atomic force microscope. The image produced by an AFM is based on the reflection of laser light off of the surface of the cantilever.

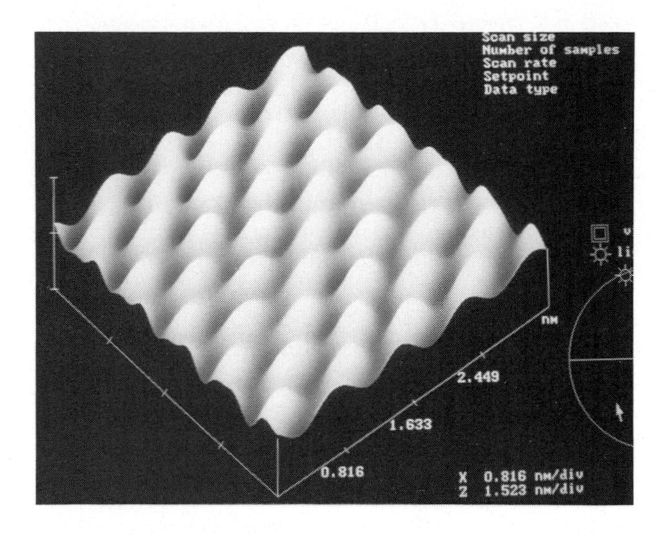

Figure 5.31 Atomic force image of mica atoms. Atomic-level resolution is evident in this photograph.

ples, the STM and AFM have enabled visualization of dislocations in alloys and lattice structure in a variety of materials. In biology, they have been used to image amino acids, proteins, and macromolecules. Many difficulties remain to be solved, however, before these new microscopes can be used fully in biology. This situation is similar to what occurred in the late 1960s with the SEM.

FURTHER READING

Barber, V. C. 1979. SEM and X-ray microanalysis of artifacts retrieved from marine archaeological excavations in Newfoundland. *Scanning Electron Microsc.* 1979(2):885–891. (Archaeology)

Baumgarten, N. June 1990. *SEM for Imaging Specimens in Their Natural State.* American Laboratory, Shelton, CT. (Environmental SEM)

Cole, E. I., C. R. Bagnell, B. G. Davies, A. M. Neacsu, W. V. Oxford, and R. H. Propst. 1988. Advanced scanning electron microscopy methods and applications to integrated circuit failure analysis. *Scanning Microsc.* 2:133–150. (Voltage constrast, semiconductor technology)

Czernuska, J. T., N. J. Long, E. D. Boyles, and P. B. Hirsch. 1990. Imaging of dislocations using backscattered electrons in a scanning electron microscope. *Philos. Mag. Letters* 62:227–232. (Electron channeling)

DiMaio, V. J. M., S. E. Dana, W. E. Taylor, and J. Ondrusek. 1987. Use of scanning electron microscopy and energy dispersive X-ray analysis (SEM-EDXA) in identification of foreign material on bullets. *J. Forensic Sciences* 32:38–47. (Forensic science)

Dingley, D. J. 1981. A comparison of diffraction techniques for the SEM. *Scanning Electron Microsc.* 1981(4):273–286, 258. (Selected-area electron channeling, electron backscattered patterns, micro-Kossel X-ray diffraction)

Doehne, E., and D. C. Stulik. 1990. Applications of the environmental scanning electron microscope to conservation science. *Scanning Microsc.* 4:275–286. (Conservation science, environmental SEM)

Gabriel, B. L. 1985. *SEM: A User's Manual for Materials Science.* American Society for Metals, Metals Park, OH. (Uses in metallurgy and materials science)

Goldstein, J. I., D. E. Newbury, P. E. Echlin, D. C. Joy, C. Fiori, and E. Lifshin. 1981. *Scanning Electron Microscopy and X-Ray Microanalysis. A Text for Biologists, Materials Scientists, and Geologists.* Plenum, New York. (General theory)

Goldstein, J. I., and H. Yankowitz. 1975. *Practical Scanning Electron Microscopy.* Plenum, New York. (General theory, electron channeling, selected-area channeling, electron beam induced-current, magnetic contrast)

Holcomb, D. N. 1991. Structure and rheology of dairy products: A compilation of references with subject and author indexes. *Food Struct.* 10:45–108. (Food science)

Holt, D. B. 1992. New directions in scanning electron microscopy cathodoluminescence microcharacterization. *Scanning Microsc.* 6:1–21. (Cathodoluminescence)

Jahanmir, J., B. G. Haggar, and J. B. Hayes. 1992. The scanning probe microscope. *Scanning Microsc.* 6:625–660. (Scanning tunneling and atomic force microscopy)

Kiss, K. 1987. Industrial problem solving with microbeam analysis. *Scanning Microsc.* 1:1515–1538. (Industrial uses of SEM and associated techniques in problem solving)

Kreindler, H. 1986-1987. The application of SEM for authentication of an important find of year five shekels of the Jewish war. *Israel Numismatic J.* 9:38–45. (Numismatics)

Laschi, R. G. P., and P. Versura. 1987. Scanning electron microscopy application in clinical research. *Scanning Microsc.* 1:1771–1795. (SEM in medical research)

Leapman, R. D., and S. B. Andrews. 1991. Analysis of directly frozen macromolecules and tissues in the field-emission STEM. *J. Microsc.* 161:3–19. (STEM imaging of biological macromolecules)

Peters, K.-R. 1985. Working at higher magnifications in scanning electron microscopy with secondary and backscattered electrons on metal coated biological specimens and imaging macromolecular cell membrane structures. *Scanning Electron Microsc.* 1985(4):1519–1544. (High-resolution imaging)

Pfefferkorn, G. E. 1984. The early days of electron microscopy. *Scanning Electron Microsc.* 1984(1):1–8. (History)

Robinson, V. N. E. 1987. Materials characterization using the backscattered electron signal in scanning electron microscopy. *Scanning Microsc.* 1:107–117. (Materials characterization)

Saglie, F. R. 1988. Scanning electron microscope and intragingival microorganisms in periodontal diseases. *Scanning Microsc.* 2:1535–1540. (Dentistry)

Schmidt, D. G. 1982. Electron microscopy of milk and milk products: Problems and possibilities. *Food Microstruct.* 1:151–165. (Food science)

Versura, P., and M. C. Maltarello. 1988. The role of scanning electron microscopy in ophthalmic science. *Scanning Microsc.* 2:1695–1723. (Ophthalmic science)

Wickramasinghe, H. K. 1989. Scanned-probe microscopes. *Sci. Am.* 261(4):98–105. (Scanning tunneling and atomic force microscopy)

Wilkinson, A. J., J. T. Czernuszka, N. J. Long, and P. B. Hirsch. 1992. Electron channelling contrast imaging in the JSM-6400 scanning electron microscopy. *JEOL News* 30(1):2–4. (Electron channeling)

Wischnitzer, S. 1981. *Introduction to Electron Microscopy.* Pergamon, New York. (General theory)

6

Specimen Preparation for TEM

Most, if not all, samples for transmission electron microscopy require some degree of preparation prior to observation in the microscope. The first factor that must be considered is the final size of the sample that can be viewed in the TEM. The standard for the TEM is a specimen support disk, called a grid, 3 mm in diameter. These grids are generally made of thin copper (about 50 μm thick) and are available with a variety of opening patterns (Figure 6.1). Grids made of gold, nickel, nylon, and other materials are available for special work such as immunolabeling or X-ray analysis.

Specimen preparation for biological samples requires special attention to the labile molecular architecture of living material. The choice of technique depends on the information sought and the type of specimen. Most techniques include the addition of some electron-dense agent to provide differential electron contrast to the electron-lucent biological sample. With this in mind, three broad approaches can be considered in preparing biological samples for TEM.

Small or thin samples can be mounted whole with the background stained (negative staining). If a sample is too thick for adequate resolu-

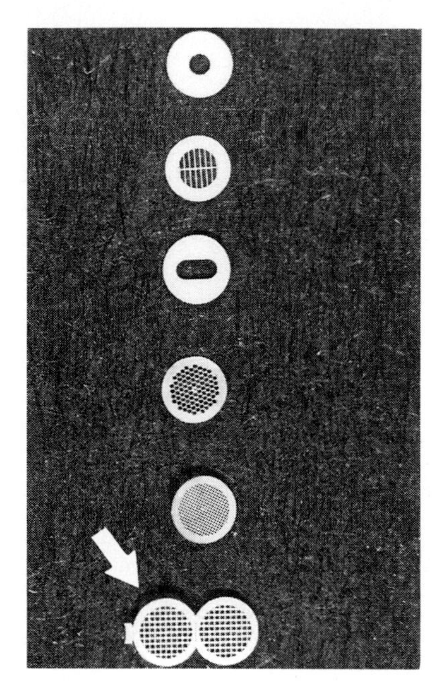

Figure 6.1 Representative styles of TEM grids. TEM grids are available in many mesh patterns. All are 3 mm in diameter. Selection of mesh size or grid bar pattern depends on the specimen preparation technique.

tion (generally more than about 200 nm), an ultrathin section of it (generally less than 100 nm) can be cut with an ultramicrotome. Finally, a replica of the sample can be made from material with known electron-optical properties. This chapter discusses the general techniques for TEM sample preparation; suggested readings at the end of the chapter provide application details.

NEGATIVE STAINING OF SMALL PARTICULATES

Negative staining is a technique that has been in use in TEM specimen preparation for over thirty years. For the visualization of small biological specimens (e.g., viruses, isolated organelles, or macromolecules), it provides a fast, high-resolution method of adding contrast to an electron-lucent biological sample. The goal of negative staining is to stain the background around the specimen with a compound that strongly scatters electrons, such as a heavy-metal salt solution. As the solution dries, it forms a glassy film on the support surface interrupted by "casts" of the electron-lucent biological material (Figure 6.2). At the edges of these casts are areas of the stain film that differ in thickness, corresponding to the surface irregularities of the specimen. These areas of different mass thickness scatter electrons differentially and provide the image contrast necessary for viewing.

Negative Stains Contain Heavy Metals

All of the materials used for negative staining contain a heavy-metal atom. To be useful as a negative stain, the material should also (1)

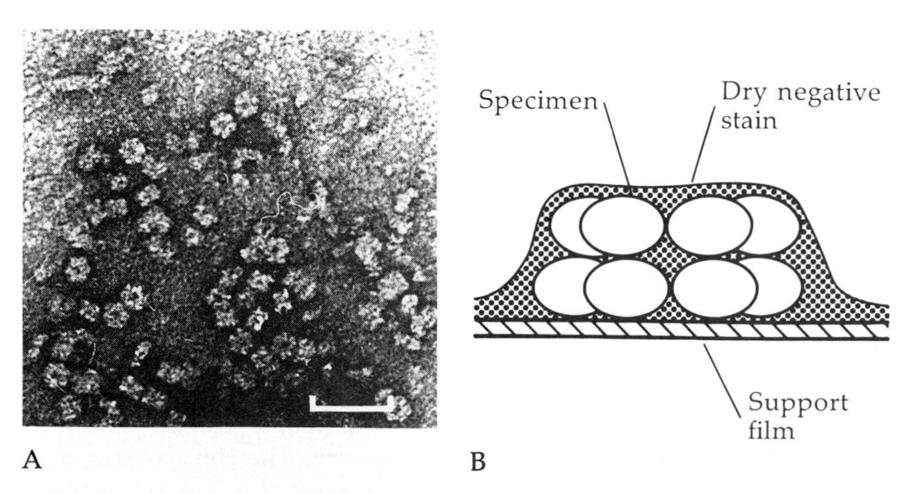

Figure 6.2 Negative staining of particulate specimens. (A) Negatively stained preparation of the enzyme glutamine synthetase. (B) Schematic cross section of the negative-staining process. The mass thickness of the stain is greatest right next to the specimen.

be highly soluble, (2) have a high melting point (beam stability), (3) have a small molecular size for good penetration, (4) dry into a smooth (glassy) film, and (5) not react with (positively stain) the specimen. A number of materials fit this description.

One of the most frequently used negative stains is uranyl acetate. This stain is also used in ultrathin-section staining (discussed later in this chapter), is useful in the acidic pH range, and is generally used as a 0.5% to saturated aqueous solution.

Another widely used negative stain is phosphotungstic acid (PTA). PTA is a good general negative stain. It is usually applied as the sodium salt at a pH of 6 to 8 and a concentration of 0.5% to 2% in water. It can tolerate relatively high concentrations of nonvolatile buffers while maintaining its effectiveness. Other stains with similar properties, such as silicotungstic acid, also give excellent results.

Ammonium molybdate, a third negative stain, in a 1% to 2% aqueous solution works at a pH of 7 to 7.4. It is especially useful in providing contrast to membrane-bounded systems such as mitochondria and other organelles.

Small Particulates Need Additional Specimen Support

To apply a negative stain to a small biological sample or to visualize a particulate materials-science sample, the sample must be supported on the microscope grid. Since the specimen is much smaller than the openings of the grid mesh, the grid must have a continuous coating. Three types of coated grids are commonly used as specimen supports for negative staining.

The clearest supports are made of plain evaporated carbon, which is deposited on a smooth surface, such as a piece of glass or mica, in a vacuum evaporator (discussed later in this chapter). After the thin carbon film is evaporated onto the support, it is scored and floated off the support onto the surface of a dish of dust-free distilled water. Bare copper grids are brought up under these squares of carbon film, and residual water is carefully blotted away. If the film is thin enough, it will remain attached to the grid with no further treatment; if it is too thick, an adhesive may be necessary. The carbon-film grids generally work better if prepared immediately prior to use.

A second type of support can be made by casting a thin plastic film. In this process a plastic, such as Formvar (polyvinyl formate), is dissolved to make a fairly weak solution (e.g., 0.25%) in an appropriate solvent, such as dichloroethane. A clean glass microscope slide is dipped into this solution and allowed to dry. The thin coating that results is floated off the slide onto the surface of a dish of distilled water. A number of bare copper grids are laid out on this floating film, and the whole thing is picked up with a screen or piece of paper. Plastic films generally adhere well to the grids and are less susceptible to cracking than are pure carbon films. However, they are also generally less stable under the beam and have higher background contrast.

A third approach in making specimen supports is to apply a light coating of carbon to a plastic film. Adding a small amount of carbon has little effect on the background contrast and significantly stabilizes the support film. Evaporation of the carbon for film generation or plastic-film stabilization is discussed later in this chapter.

Three Common Negative-Staining Procedures

There are a number of methods for preparing negatively stained samples. Since none of them is especially laborious, it is often advantageous to try more than one for a given sample to find which method works best.

The simplest method is the drop method. In this technique, the specimen (dissolved in an appropriate vehicle at a concentration generally less than 0.1 mg/mL) is placed on a coated grid and allowed to adsorb for about a minute. The bulk of the sample is then blotted off the edge of the drop with filter paper. A drop of stain solution is added and allowed to incubate for 30 seconds, after which the grid is blotted off from the side. The specimen is then ready for viewing.

Rather than placing the specimen on the grid, an alternative method is to place the grid or support film on the surface of a small volume of the specimen mixture and allow the specimen particles to adsorb for several minutes. The grid or support film is then transferred to the negative stain. In this method, unbound materials diffuse into the stain droplet. The grid with adhering particles is then picked up and blotted dry from the edge.

A third approach is to mix equal volumes of the specimen preparation with a negative stain and spray them at the coated grid with a nebulizer or modified airbrush. The resulting fine droplets collide with the grid surface, making small areas with the proper stain thickness. With the modified airbrush, the total sample solution required is less than 20 μl, making it an especially useful tool when only small samples are available. With either spray technique, care should be exercised with the aerosol dispersion of the heavy metal stains and with hazardous specimens.

ULTRATHIN SECTIONING OF LARGER SAMPLES

Most biological specimens have a minimum dimension of more than 100 nm, but a thinner subsample must be used for TEM. Ultrathin sectioning is the most widely used sample preparation technique in the biological sciences. The purpose of ultrathin sectioning is to render a representative plane of the specimen thin enough to be viewed in TEM. Although ultrathin sectioning is the preparation technique that makes sample observation in the TEM possible, it is actually conducted in the midst of a number of preparation steps, each highly dependent on the successful completion of the preceding step.

The following outline illustrates the position of the sectioning process in the flow of routine biological sample preparation:

1. Specimen isolation
2. Killing and fixing
3. Dehydration
4. Resin infiltration and embedding
5. Ultrathin sectioning
6. Staining
7. Interpretation

Specimen Isolation Is Often the Most Critical Step

The degradation of a biological specimen starts, at the latest, with excision of a sample from the whole organism. Thus, the time between removing a sample and fixing it should be as short as possible. Animal tissue is often more susceptible to observable artifacts induced by isolation trauma than is plant material. In many cases, observable ultrastructural changes occur so rapidly post mortem that they would obscure or alter the phenomena under investigation if the tissue were removed directly from the animal. In these cases, in-situ primary fixation via vascular perfusion (discussed later in this chapter) provides an alternative. Even this technique sometimes falls short of preventing undesirable changes during sample isolation, however, and other approaches (also discussed later in this chapter) are required. With any biological sample, transfer to the fixation fluids should take place well before there is any chance of damage from desiccation. This timely transfer can be accomplished by removing a sample larger than is required, immersing it in buffer or fixation solution, and subsampling from its interior. Diffusion rates play an important part in the speed of chemical fixation. Fixing speed and quality are thus related to the surface-to-volume ratio of the specimen. Limiting the maximum final dimension of the sample to 1 mm is generally a good practice for initial fixation of excised tissue and secondary fixation of perfused tissue. This maximum size allows some small whole plants and animals or at least whole organs to be fixed. Plant root tips, for example, often have diameters less than 1 mm, and simply cutting them from the plant produces an adequate sample size. For the physical fixation methods discussed in the following sections, even smaller samples are required.

Killing and Fixing Are the Next Steps

Ideally, the image seen in a TEM should reflect the organization of cellular components, to the molecular level, that was present just before sample preparation began. Since life is defined by the arrangement and condition of the same molecules, the role of killing and fixing is to stabilize them in lifelike condition. Since an exactly lifelike condition would mean that the tissue was alive, however, a series of compromises in structure must be made for the sake of stability. Properly killed and fixed tissue should balance the compromises to yield the most information of interest.

Tissue may be fixed by physical or chemical means. The method of physically fixing TEM samples that produces the highest fidelity is ultrarapid freezing, an approach developed initially for freeze-fracture preparations but also useful for freeze-substitution preparations. The most common fixation method is a two-stage chemical regimen using aldehyde and organometallic crosslinking. Although chemical fixation is somewhat easier to accomplish and does not require as small a sample, ultrarapid freezing has decided advantages when ephemeral cellular events are to be studied or when highly labile histochemical or immunological characteristics are to be observed.

The first requirement of fixation is to stabilize (fix) the cellular components in place and in as close to a lifelike condition as possible. If histochemical or immunological labeling is to be done, this lifelike condition must persist to a molecular level on the targeted reaction or antigenic sites (although uninvolved ultrastructural details may be sacrificed in these investigations).

A second major function of fixation is to protect the sample from disruption, dimensional change, and loss of material during subsequent processing steps. Samples that will undergo ultrathin sectioning need to be embedded and held in a supporting matrix and are subjected to a series of highly extracting solvents in the embedding process.

A final role of fixation is to prepare the sample material for exposure to the electron beam. Even after undergoing ultrathin sectioning, the fixed sample must still be able to withstand irradiation by a multikilovolt electron beam without disintegrating rapidly.

Since the killing and fixing procedure always involves some compromises with the living state and introduces some artifacts, judging the quality of the fixation is problematic. A number of general guidelines, however, can be followed to determine the quality:

1. Cell membranes should be continuous and smooth.
2. The cytoplasm should have no empty spaces (final appearance agrees with light microscopy of similar living material).
3. Loss of cellular material by extraction should be as little as possible.
4. Preservation of the structure by different means should agree.

Chemical Fixation To accomplish a good chemical fixation, the fixative must meet several requirements. First, a fixative should work quickly to arrest the life functions of a cell. A cell that has been stopped slowly will have ultrastructural alterations. In the extreme case, fixation could take place so slowly that the endogenous enzymes of a cell could start to digest the cell (a process called autolysis) during fixation. A second requirement closely related to the first is that the fixative preserve the fine detail of the cell without rearranging the components. Third, the reactions between the various classes of biomolecules and the fixative should be known to some degree to interpret the resultant image correctly. If possible, the fixation procedure should be relatively economical and safe. Finally, the fixative should have a sufficient fixation capacity at usable concentrations. Normally about a 5:1 or greater fixative-to-tissue volume ratio seems to work best. No single fixative tried to date has all of these attributes, and the answer to the problem lies in multiple fixation steps.

Primary Fixation Most biological material is fixed by at least a two-stage process. The two steps are aimed at fixing different classes of biomolecules, and using more than a single fixative generally produces a synergistic result. Usually the first step is to fix the sample in an aldehyde or mixture of aldehydes because aldehydes penetrate more deeply than the metallic oxide fixatives do. Three aldehydes are used commonly for TEM sample preparation.

Glutaraldehyde is the most common. Much of its efficiency as a fixative seems to stem from the fact that it is a dialdehyde (O=C–C–C–C=O). The chemistry of fixation of biomolecules by glutaraldehyde is complex. Reactions between glutaraldehyde and proteins seem to involve the formation of crosslinked complexes of pyridine-derived polymers formed from both soluble and membrane-bounded proteins, which may account for the high degree of spatial integrity in glutaraldehyde-fixed tissues.

The reactions of glutaraldehyde with proteins release large quantities of hydrogen ions into the environment and require oxygen (O_2) from it. Hypoxia and drop in pH along the reaction front may explain the slow penetration sometimes reported with glutaraldehyde fixation. Thus, optimum fixation usually requires some pH buffer system and may benefit from aeration or other oxygen sources, e.g., hydrogen peroxide (H_2O_2). Although glutaraldehyde forms insoluble complexes with many cell proteins, the cells remain osmotically active, and adding tonicity-balancing solutes is usually recommended. The pH buffer often can be adjusted to perform this function.

Less information exists about the reactions of glutaraldehyde with other biomolecules in the cell. Glutaraldehyde appears to stabilize and perhaps bind to glycogen. Lipids are not stabilized by glutaraldehyde. If cells are fixed only with glutaraldehyde, over 90% of the lipids in the cell can be extracted during subsequent dehydration steps.

Glutaraldehyde generally is available as an 8%, 25%, or 50% solution in water. Nonreactive polymers may form during storage, especially at warmer temperatures, so glutaraldehyde should be stored at less than 4°C for a longer shelf life. Although commercially available grades are adequate for most TEM applications, for critical work, glutaraldehyde can be redistilled to free it from polymers and checked spectrographically. Working buffered solutions should be made just before use and normally contain from 0.5% to 10% glutaraldehyde by volume.

Pure formaldehyde (H–C=O), the smallest of the aldehydes, is a gas at room temperature. It is the second most commonly used aldehyde in TEM specimen fixation. Although its fixing reactions involve chiefly proteins, formaldehyde also fixes nucleic acids. The protein reaction products appear to be quite different from those of glutaraldehyde fixation reactions. Fewer large, crosslinked structures are formed, and more material is subject to leaching, especially if the fixation takes place over a long period of time. Formaldehyde does not fix lipids or polysaccharides well. In spite of these shortcomings, formaldehyde does penetrate the sample rapidly, which may outweigh some of its deleterious effects, especially in large or dense samples. As discussed later in this chapter, it is often used simultaneously with other aldehydes, normally in the 1% to 10% range in a buffered medium.

Commercially available formaldehyde solutions (formalin) generally have about 15% methanol included to stabilize them. The presence of methanol has been shown to have a highly damaging effect on tissues being fixed. For TEM applications, formaldehyde is usually prepared by dissolving some of its polymer, paraformaldehyde, in hot water shortly before use. Often the pH is raised by adding sodium hydroxide (NaOH) to aid in the dissolution and then lowered back to physiological levels with hydrochloric acid (HCl). If these steps are taken, attention should also be paid to the osmotic effect of the resultant salt, sodium chloride (NaCl), and the highly carcinogenic reaction products of formaldehyde and HCl vapors.

Acrolein is the third aldehyde used with any frequency in TEM preparations. Also known as acrylic aldehyde (CH_2=CH–CHO), acrolein is a very rapidly penetrating aldehyde useful for the primary fixation of large samples and those with especially dense cell walls. It is rather toxic, has a highly irritating odor (it has been used in gas mixtures in warfare), polymerizes easily (especially under light), and is flammable. Histologically it is also the most extractive of the aldehydes, removing lipids extensively. Acrolein also destroys most enzyme activity. It is usually used at concentrations of 1% to 10% in combination with other aldehydes and buffers, to attempt to speed up initial fixation.

As mentioned earlier, aldehydes are often used in mixtures to ensure the best combination of fixative characteristics. For example, the slow penetration of glutaraldehyde can be counteracted by mixing it with another aldehyde that penetrates faster but may not fix as well. The most common combination is a mixture of glutaraldehyde and formaldehyde. Originally it was thought that the more rapidly penetrating formaldehyde temporarily stabilized structures that were then more permanently crosslinked by glutaraldehyde. More current data, however, support the idea that formaldehyde changes the overall fixation chemistry, and some of the benefit may be from altering the O_2 requirement of glutaraldehyde fixation and/or the reaction products formed.

Primary fixation steps are generally those most sensitive to temperature during fixation. Low fixation temperatures, usually in the range of 0°C to 4°C, are often used in an attempt to avoid autolytic effects. Since the fixation reactions are also subject to thermodynamic principles, however, a better approach is to start fixation at ambient or physiological temperatures. One of the most notable effects of low-temperature fixation is the loss of microtubules. In some plant tissues, more extraction occurs when fixation is started at a lower temperature.

Secondary Fixation As discussed earlier, aldehyde fixation leaves major classes of biomolecules unfixed. Indeed, without proper consideration of fixation time and temperature, large quantities of cellular material may be leached out during primary fixation. For general morphological work, the usual approach is to follow up a one-hour to overnight aldehyde primary fixation with a complementary secondary fixative, of which the most common is osmium tetroxide (OsO_4).

Osmium tetroxide is a highly volatile crystalline solid that dissolves slowly in water. Its main mode of action is to crosslink lipids or other molecules at points of nonsaturation. It also appears to react with (bind to) some structures fixed by aldehydes. OsO_4 alone was the first commonly used fixative for TEM sample preparation, but it has become less favored as the primary fixative because of its low penetration rate. If OsO_4 is used as a primary fixative, the required sample size is less than 0.5 mm in its maximum dimension. Even with samples this small, unfixed material can exist in the center. OsO_4 works well as the secondary fixative (after aldehyde fixation has been accomplished), fixing the remaining cellular components well enough that little leaching of cell contents occurs during subsequent sample preparation steps. The specimen-bound Os (atomic number 76) in OsO_4 also effectively scatters electrons, thus acting as a general electron stain, and greatly increasing the specimen contrast. As a result of fixation with osmium, cell membranes lose their osmotic properties and are generally less environmentally sensitive in subsequent processing steps, so buffers are generally not needed.

Osmium tetroxide is usually added to the tissue to be fixed at a concentration 0.5% to 2% in an aqueous solution. The fixation reaction proceeds at a rapid rate at room temperature, and complete fixation usually takes no more than two hours with no apparent difference from low-temperature fixation, other than rate.

Since OsO_4 is a very volatile compound in both solid and solution form, it should always be used under a fume hood because it has a choking odor and can fix the epithelial cells of the eyes and nasal membranes quickly. OsO_4 is generally prepared from the solid as a stock solution, which is stable for weeks if kept in clean glassware, at low temperature (4°C), in the dark, and uncontaminated.

An alternative general-purpose secondary fixative is ruthenium tetroxide (RuO_4). Its fixative properties are closely related to those of OsO_4, but it is reported to be less toxic and less volatile than OsO_4 and is also a stronger oxidizing agent. In addition to straight-chain points of unsaturation, it also reacts more strongly with aromatic rings than OsO_4 does, and it reacts with proteins, glycogen, and monosaccharides. RuO_4 has been introduced as a potential replacement for OsO_4 becuase of its lower cost per sample. Although RuO_4 is claimed to be less toxic than OsO_4, it is still quite volatile. In light of this characteristic and its apparently higher reactivity than that of OsO_4, RuO_4 should be handled with the same precautions as OsO_4. In many tissues RuO_4 seems to react to give more intense structural staining than OsO_4 does, which may be due to the faster reaction rate or the wider range of cellular chemicals with which RuO_4 reacts. As with OsO_4, RuO_4 is generally added to the sample in an aqueous solution, although at a lower concentration (e.g., 0.1%). Its quick reaction with some organic buffers (e.g., PIPES) should be taken into consideration at the outset.

Occasionally other fixation agents are useful for TEM samples. Potassium permanganate ($KMnO_4$) and other permanganates, such as $NaNnO_4$ and $LaMnO_4$, are rapidly penetrating, oxidative fixatives that are useful in membrane studies or when thick cell walls must be penetrated (e.g., in yeasts). Because of their penetrating power, permanganates were used more extensively before glutaraldehyde fixation became popular. Permanganates fix lipids and phospholipids well and also fix glycogen to some extent. They are highly reactive, making attention to treatment time important. Permanganates tend to destroy much of the rest of the cell ultrastructure, however, so their applications are limited (Figure 6.3).

Some investigators introduce a third step, fixation with uranyl acetate, after the secondary fixation. This additional fixation stabilizes membranes and nuclear material, adds to the overall contrast of the sample, and makes staining with uranyl acetate after sectioning unnecessary. Usually a 0.5% to 2% solution is used for about 15 minutes, followed by several water rinses.

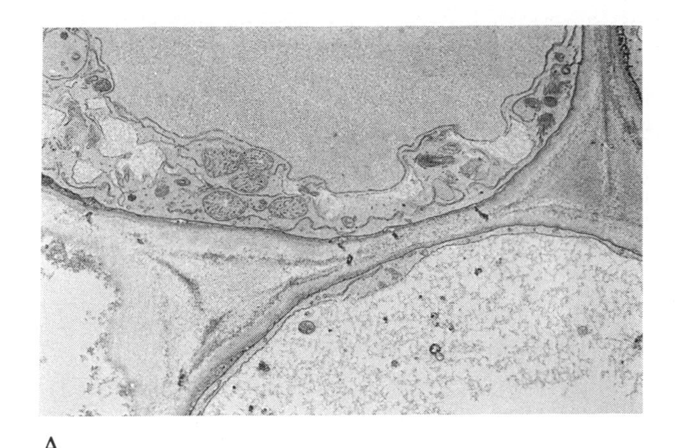
A

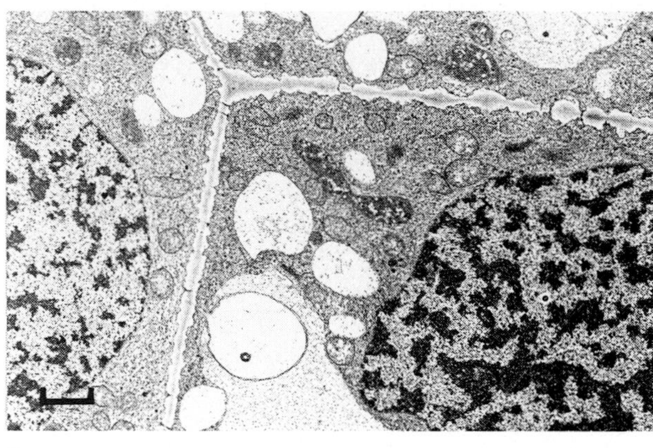
B

Figure 6.3 Comparison of chemical fixation methods. (A) Ultrathin section of an onion root cell fixed in $KMnO_4$. (B) Ultrathin section of onion root cell fixed with glutaraldehyde and OsO_4. Permanganate fixation yields distinct membranes but little else. Fixation is much more uniform with glutaraldehyde/osmium tetroxide fixation. Bar = 1 μm.

Fixation Media The reaction medium for chemical fixation is an important factor in producing a fixed image free of uncontrolled artifacts. Two of the most important characteristics of the medium are pH and osmotic potential. The pH (and sometimes tonicity) is usually controlled by a buffer system, which is especially important in the primary-fixation stage, when the cell membranes are osmotically active and many cellular components are subject to leaching. Buffered solutions are also used to remove the unreacted aldehyde fixatives from the sample and its surrounding medium prior to secondary fixation. They may also be used in the secondary-fixation steps.

The pH buffer, by definition, should resist changes in pH during fixation. The buffer is especially important during glutaraldehyde fixation because of the number of hydrogen ions liberated in the crosslinking reaction. The buffer should, as much as possible, not react with the fixative agent (or other added chemicals, if present). The buffer itself should not extract cellular components. Ideally, the buffer also should not be acutely toxic to the material being fixed (or to the person fixing the material). Many types of pH buffers are used in TEM sample preparation. Within a given class, the choice is perhaps best made based on the assumed (or measured) pH of the in-vivo system and the pK_a (dissociation equilibrium constant) of the buffer to be used. For maximum buffering capacity, the pK_a should be near the pH of the system being buffered.

Phosphate buffers are used widely for general TEM preparations, since they fulfill most of these requirements adequately. These buffers are most often prepared from stock solutions of sodium hydrogen phosphate ($NaHPO_4$) and sodium dihydrogen phosphate (NaH_2PO_4) at concentrations of 0.01 M to 0.2 M. Although the osmolarity of the buffer can, to a degree, be adjusted by the molarity, sucrose or sodium chloride (NaCl) are commonly used to maintain a given osmotic poten-

TABLE 6.1
Common Buffers Used in EM and Their pK_a Values
Buffers have an effective range of about 2 pH units on either side of their pK_a value.

BUFFER	pK_a
Phosphoric acid	2.12 (pK_{a1})
	7.21 (pK_{a2})
	12.32 (pK_{a3})
Acetic acid	4.75
MES	6.15
Cacodylic acid	6.19
PIPES	6.80
MOPS	7.20
HEPES	7.55
Barbital (Veronal)	7.43
TRIS	8.30

Buffers listed by acronym are from Good, N. E., G. D. Winget, W. Winter, T. N. Connolly, S. Innawa, and R. M. M. Singh. 1966. Hydrogen ion buffers for biological research. Biochemistry *5:467–477.*

tial when raising the concentration of phosphate ions might lead to precipitation.

Phosphate buffers can be prepared from stock solutions of mono- and dibasic phosphates to cover the pH range of 6 to 8 (although they are feeble at pH $\geq$ 7.5). Although nontoxic and inexpensive to prepare, phosphate buffers tend to be more extractive than some of the other buffers described here, can precipitate or bind ions less soluble than sodium, and easily become contaminated with microorganisms.

Cacodylate (or dimethyl arsenic acid), another widely used pH buffer, is usually dissolved in water to make a 0.1-M to 0.2-M solution that is then adjusted to the desired pH with hydrochloric acid (HCl). Cacodylate buffers are generally regarded as less extractive than phosphate, although this may not be true with some samples. Cacodylate buffers do not support microorganism growth because of the arsenic component. They are quite stable when refrigerated. In addition, they do not precipitate many of the divalent ionic species (notably calcium and magnesium) that can be precipitated by phosphate buffers.

A third class of buffers is the zwitterionic buffers (also known as Good's buffers), a series of complex organic acids that are dipolar ions (Table 6.1). They are relatively nontoxic, do not react appreciably with mineral cations, cover the usual range of biological samples, and are stable. These buffers are excluded by unfixed biological membranes, and thus buffer the medium surrounding the cells but not the intracellular contents (at least initially). In some systems they are reported to cause less cellular leaching after glutaraldehyde fixation than phosphate buffers do.

In addition to being buffered, the fixation medium needs to be near osmotic equilibrium with the sample material to avoid gross distortion on an ultrastructural level. The equilibrium often is obtained by adding sucrose or neutral salts to the pH buffer medium. Because glutaraldehyde and formaldehyde, at the concentrations generally used, can significantly change the osmotic potential of the buffered fixative, their effect should be considered.

Cryogenic Fixation Techniques Work Faster Than Chemical Means

Although chemical fixation is the most common method of stabilizing biological samples for TEM sample preparation, there are considerable drawbacks with even the best chemical-fixation regimen. First, all chemical fixatives in use at present must diffuse across the cell membrane prior to the onset of fixation, and the fastest they can move through the sample is only a few micrometers per second. The delay becomes a significant consideration when trying to study ephemeral cell activities.

A second consideration is that chemical methods fix different biomolecules at different rates within the cell. In some plant cells, cytoplasmic streaming has been seen for several minutes after starting fixation, even in mixtures containing acrolein and glutaraldehyde. The cell activity grinds to a halt rather than being frozen in time, causing artifactual displacements within the cell. In studies requiring the localization of unbound ions, chemical fixation is of little use, since the common fixatives do not stabilize ions at their native sites. For some immunological and histochemical labeling as well, the crosslinking nature of the aldehyde and OsO_4 fixatives can destroy the activity of the target molecules.

Ultrarapid freezing is a physical alternative to chemical fixation. With ultrarapid freezing, the contents of a cell are frozen so quickly that any ice crystals formed are below either the resolution limit either of the microscope or of the subsequent sample preparation techniques.

There are four approaches to ultrarapid freezing in general use. In addition, there is a fifth cryogenic method, high-pressure freezing, that gives qualitatively similar results to ultrarapid freezing by hyperbarically changing the crystallizing rate of cellular water. To freeze biological materials ultrarapidly at ambient pressures, a cooling rate in the range of 10^{4}°C/s to 10^{5}°C/s is necessary to suppress the formation of ice crystals. At the pressures employed by the high-pressure freezer, however, the required cooling rate is reduced to about 100°C/s.

The first method of ultrarapid freezing was plunge freezing (Figure 6.4A). In this method, a small sample is rapidly immersed in a cryogen, usually liquid propane or freon, held near its freezing point in a bath of liquid nitrogen. Although colder than liquid propane or freon, liquid nitrogen is at its boiling point at atmospheric pressure. The introduction of a relatively warm specimen results in the formation of an insulating envelope of gaseous nitrogen boiling around the specimen, which dramatically slows the freezing rate.

After plunging, the excess cryogen can be wicked away quickly with a bit of filter paper and the specimen stored under liquid nitrogen. This simplest and cheapest method of freezing is limited by the thermal transfer characteristics of the specimen and the specimen support. With most materials it is difficult to freeze more than a superficial layer, considerably less than 10 μm, ultrarapidly. The depth of frozen material in a sample can be increased by adding cryoprotectant materials such as glycerol or dimethyl sulfoxide (Figure 6.5), which lower the freezing point and raise the recrystallization point. The additives themselves, however, can produce artifacts that require chemical fixation to prevent, thereby nullifying one of the major reasons for using cryofixation.

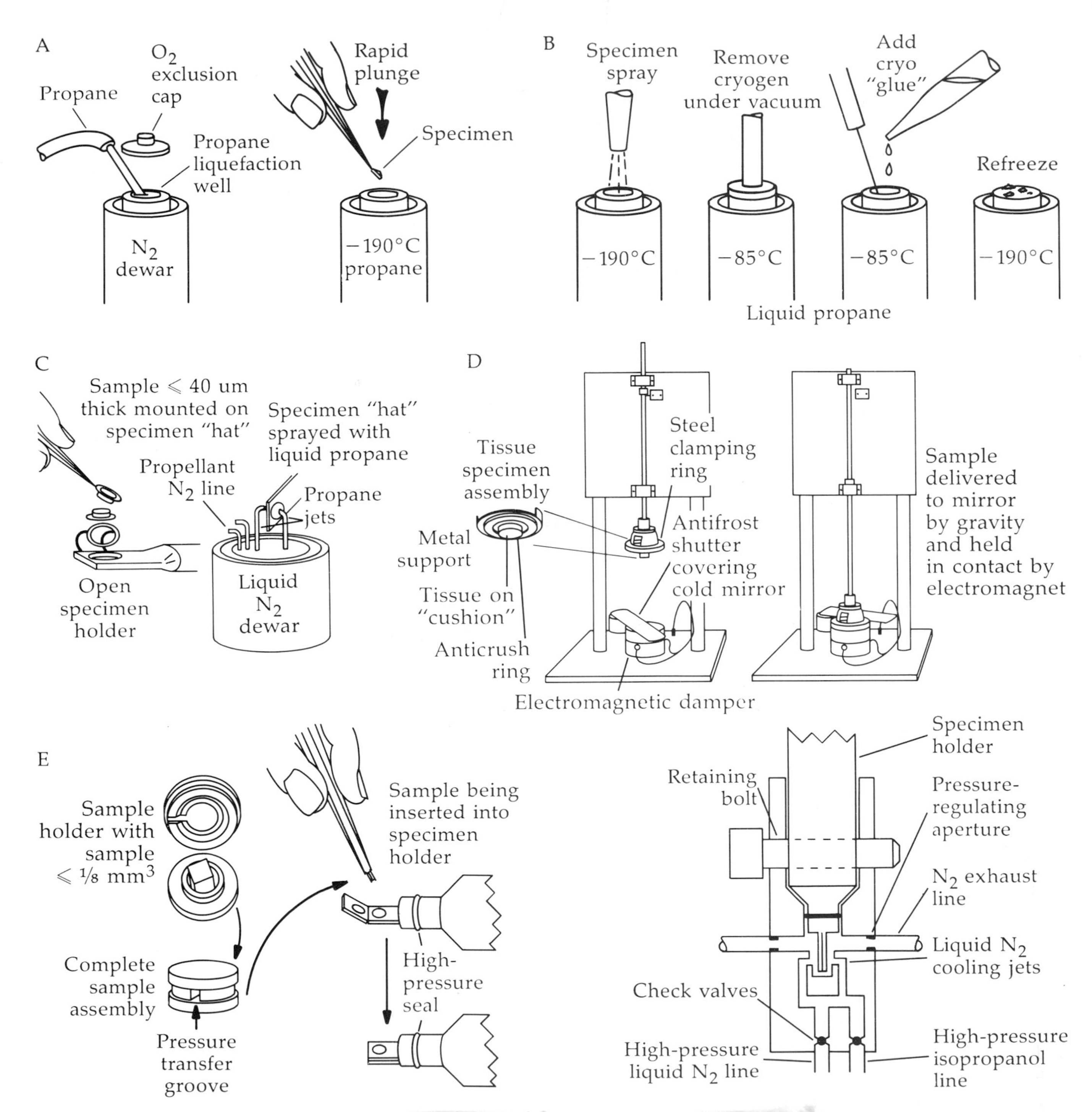

Figure 6.4 Cryogenic fixation methods. (A) Plunge freezing: The sample is mounted on a support and plunged rapidly into a liquid cryogen. (B) Spray freezing: The sample is plunged into the cryogen as an aerosol. The cryogen is then removed with a vacuum, leaving the frozen sample as a powder, which is made into a paste with a cryogenic glue for subsequent handling. (C) Propane-jet freezing: The sample is held between two thermally conductive supports, and liquid propane is sprayed at both sides, increasing the heat flux and the thickness of the sample that is well frozen. (D) Metal-mirror ("slam") freezing allows larger tissues to be frozen ultrarapidly by rapidly drawing heat away through a polished metal surface that has been cooled to cryogenic temperatures. (E) High-pressure freezing represents a tenfold improvement, in terms of the thickness of the sample that can be frozen, over ultrarapid freezing, which is limited by the thermodynamic properties of ice to a sample thickness of about 40 μm. At present, high-pressure freezing requires an elaborate, hydraulically pressurized liquid-nitrogen-jet freezer.

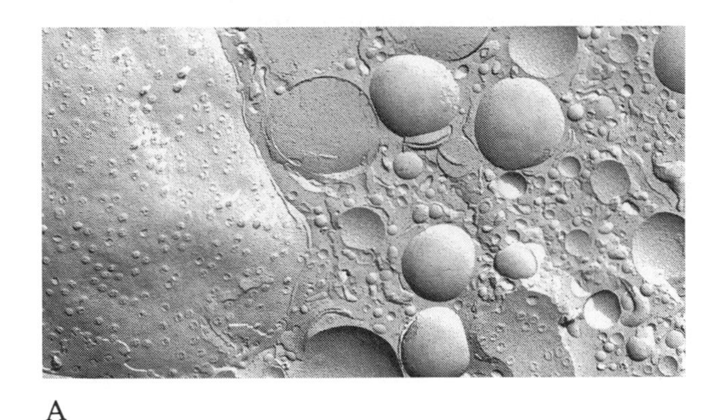

Figure 6.5 Cryoprotection of a mouse liver cell. (A) Freeze-fracture replica of a mouse liver cell incubated in 30% glycerol prior to plunge freezing. Glycerol disrupts membrane systems in unfixed tissue. (B) Replica from tissue from the same liver fixed with 2% glutaraldehyde prior to glycerol incubation. In this treatment the laminar structure of the endoplasmic reticulum is preserved. Bar = 1 μm.

A second approach, developed early in the history of TEM specimen preparation, is spray freezing (Figure 6.4B). In this method, the specimen is divided into a fine spray by an airbrush and sprayed onto the surface of a container of liquid propane. The propane is then drawn off the container with a vacuum, leaving a frozen powder of sample droplets that can be made into a paste with a cryogenic binding agent such as butylbenzine and solidified at liquid nitrogen temperatures. This system works only with specimens that can be divided into an aerosol, and in addition, its use is limited to those specimens that are not harmed by the high shearing forces of the spray formation process. If the specimen can withstand the forces involved, the high surface-to-volume ratio of the spray droplets allows very high rates of heat transfer, and the specimens are frozen beautifully.

Cold-metal-block freezing (or "slam" freezing) is a third approach to freezing samples (Figure 6.4D). In this method, the specimen is rapidly brought into fixed and continuous contact with a very cold, polished, metal surface. The details of the contacting mechanism vary with manufacturer, but the most important characteristic is controlled, bounce-free contact (see Figure 6.4). Heat flows out of the specimen and into a metal block made of a highly conductive material, such as copper, faster than it could through a cryogenic liquid. If proper contact is made, the surface layer of the sample can be frozen ultrarapidly to a depth of 10 μm to 15 μm using liquid nitrogen as the cryogen. In theory, ultrarapid freezing with liquid helium to cool the block to about 18K should increase the freezing depth by 30% to 40%, but this benefit must be weighed against the extra cost and trouble of dealing with the liquid helium. For large, flat pieces of tissue, this is probably the best method of ultrarapid freezing at ambient pressures. It is also the only method that is really compatible with experiments involving the stimulation of specimen tissue just prior to fixation. Cold-metal-block freezing can also be adapted to portable units that allow in-situ freezing, further expanding the types of tissues that can be analyzed.

Propane-jet freezing is a fourth method of ultrarapid freezing (Figure 6.4C). Propane-jet freezers spray a stream of liquid cryogen (usually propane at about –190°C) at either side of the isolated specimen. The advantage to this mode of operation is that heat is extracted from both sides of the specimen at the same time. If the specimen is thin enough (generally ≤ 40 μm, but sometimes thicker, especially with some plant material), the whole specimen can be frozen ultrarapidly. If the specimen is much thicker, the effect is the same as spraying from one side, and the depth of ultrarapid freezing is about 10 μm to 15 μm. Since propane-jet samples are usually mounted on 3-mm supports, the lateral dimensions are, accordingly, smaller than those possible with cold-metal-block freezing. The ability to freeze thicker specimens, however, makes propane-jet freezers quite useful for many samples.

A final approach to specimen freezing is to alter the physical nature of the freezing process by increasing the pressure (Figure 6.4E). If the pressure is raised to about 2,100 atm, the freezing point of water is depressed to -22°C, and the growth rate of ice crystals is greatly reduced. Under these conditions, the rate of freezing required for crystal-free freezing is reduced to the range of 100°C/s, allowing a much larger tissue sample to be properly frozen. The process involves a device much like a propane-jet freezer that raises the pressure of a chamber surrounding the specimen rapidly to about 2,100 atm before spraying the specimen with jets of liquid nitrogen (which can be used as a cryogen at these pressures) (see Figure 6.4).

Regardless of the method of ultrarapid freezing, the subsequent handling of the specimen has a major effect on the quality of the final TEM image. One of the main considerations after ultrarapid freezing is the storage of the sample prior to the next preparation step. In general, specimens can be stored under liquid nitrogen between freezing and subsequent preparation steps. Some investigators have found that the specimens tend to become brittle after prolonged storage under liquid nitrogen. The reasons for this phenomenon are obscure. A far more common problem seems to be the warming of the specimen above the point at which solid-transition ice crystals form. In most biological material, this begins to occur rapidly above –80°C and can be induced by slow transfer from the freezing apparatus to the liquid nitrogen holding bath, poor contact with cold handling surfaces, and contact with insufficiently cooled equipment.

After been successful ultrarapid freezing, the specimen can be prepared in one of two general ways. It can be surface replicated (freeze-fracture/etch) or ultrathin sectioned (cryosectioning or freeze-substitution). Freeze-fracture is discussed later in this chapter. With cryosectioning, the specimen is mounted in a cryo-ultramicrotome, and thin, frozen sections are made and observed at very low temperatures using a cryogenic microscope stage. Freeze-substitution is more common and yields specimens embedded for ultrathin sectioning, much as chemical fixation methods do.

Specimens Need Dehydration before Embedding

Regardless of the method of fixation, if the sample is to be ultrathin sectioned, it must be transferred from a hydrated to an anhydrous state prior to infiltration with most of the epoxy embedding media. Although other types of embedding media tolerate considerable amounts of included water, they are usually employed in immunochemical or histochemical studies in which the chemical properties of the epoxy resins interfere with the desired reactions.

Freeze-Substitution

Freeze-substitution is a method of sample preparation that can combine secondary fixation with specimen dehydration. The frozen sample is immersed in a bath of solvent (such as methanol, ethanol, or acetone) at less than –80°C. After immersion, it is stored for about three days, during which time the ice molecules are exchanged with the cold solvent molecules without liquefying the water (Figure 6.6). For improved fixation, osmium tetroxide (OsO_4) may be dissolved in the solvent to stabilize structures after substitution takes place. After substitution, the temperature of the specimen can be raised, and the sample can be given several changes of plain dehydrant. The specimen is then ready for infiltration with embedding medium.

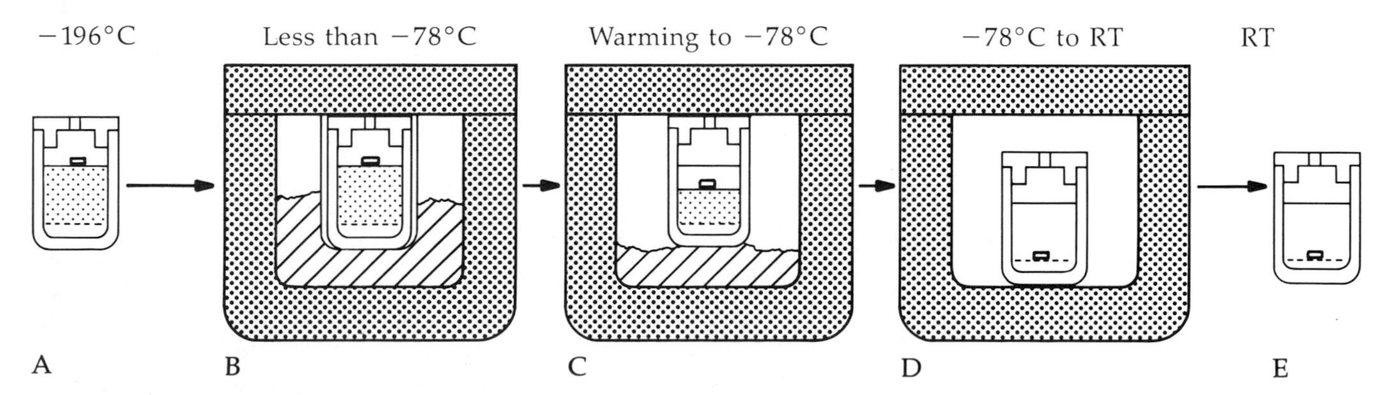

Figure 6.6 Simple apparatus for freeze-substitution. (A) A vented substitution vial with the specimen stored at liquid-nitrogen temperature. (B) The vial, filled with frozen substitution medium and specimen, is moved to dry-ice temperature. (C) As the substitution medium melts, substitution occurs. (D) The sample is allowed to warm to room temperature. (E) The substituted specimen is given several changes of pure solvent prior to infiltration with embedding medium. RT: room temperature.

Conventional (Liquid-Phase) Specimen Dehydration

With material that has not been prepared by freeze-substitution (i.e., chemically fixed), dehydration is usually accomplished by serially exchanging the water in the system with a solvent that is soluble with the embedding medium and water. Ethanol and acetone are the two most common dehydrating agents.

Once the specimen tissue has been completely fixed with glutaraldehyde and osmium tetroxide and the excess secondary fixative has been removed, dehydration generally is accomplished by transferring the specimen into a graded series of solutions containing the solvent and water. With most biological materials, this series can be fairly short, with increments in concentration such as 25%, although some fragile systems may require more gentle treatment, such as 10% steps. Usually two or three final changes in anhydrous solvent are made to assure that all traces of water have been removed.

Although both acetone and ethanol are suitable dehydrating agents, for TEM, ethanol is thought to extract more lipids than acetone does. In addition, acetone appears to cause less tissue shrinkage. It is somewhat more expensive and hydrophilic than ethanol and requires careful attention to keep the final steps dry.

Another method of dehydrating tissues is chemical, by the use of the hemiacetals 2,2-dimethoxypropane or 2,2-diethoxypropane. These reagents combine with water to yield methanol and acetone (in the case of 2,2-dimethoxypropane) or ethanol and acetone (in the case of 2,2-diethoxypropane). To improve infiltration of the sample, some investigators have made the final dehydration step a transfer from ethanol or acetone to propylene oxide, which is miscible with and tolerated by a wider range of embedding media than is alcohol. Care should be exercised with propylene oxide because it is hazardous and has a strong extracting potential.

Resin Infiltration and Embedding

The purpose of embedding the tissue in resin is to provide support during ultramicrotomy and to retain the spatial organization of the specimen section on the TEM grid. All TEM embedding media are polymers and are linked in their final polymeric form in the sample, prior to sectioning.

There are several characteristics of an embedding medium that are highly desirable. The unpolymerized medium should be soluble in a solvent that is relatively nondestructive to the specimen (e.g., acetone or ethanol). The medium itself should also be nondestructive and should not leach out the fixed tissue within the time required for embedding. Ideally the medium should be of sufficiently low viscosity that it can be infused into the sample in a reasonable amount of time, and the viscosity should not increase appreciably during the preparation time (i.e., it should have sufficient "pot" life).

Cutting properties are another major consideration. The plasticity, elasticity, hardness, and shearing strength of the polymerized embedding material are all factors, largely determined by the chemical composition of the medium, that influence the outcome of the sectioning process and have been worked out, mostly empirically, for the media in general use. Differences in the ability to stain the resulting thin sections make compatibility between the staining system and the embedding medium also important. Often different levels of these characteristics can be juggled to achieve the best medium for the task at hand. In addition, the embedding material must be resistant enough to electron-beam bombardment to withstand observation in the microscope.

Epoxy Resins For most TEM sample embedding, epoxy resins are used. As a class, these compounds offer far superior beam stability and less shrinkage than do the acrylics (discussed in the next section). Most, if not all, of the epoxies used in TEM originally were developed as commercial adhesives or polymers, and their utility as embedding media was discovered as a secondary use.

Epoxy resin mixtures used in TEM sample preparation are generally blends of four functional classes of chemicals. These are the resin, hardener, modifier, and accelerator, which are mixed together in specific ratios prior to infiltration and embedding. The epoxy resin is the main structural constituent that bears the terminal epoxy groups necessary to the crosslinking. A wide variety of epoxy resins is available to the electron microscopist, and many formulations have been tried. Three of the most widely used are Epon 812, Araldite, and vinyl cyclohexane dioxide (VCD).

Until discontinuation of its manufacture by Shell Chemical Company, Epon 812 was perhaps the most widely used epoxy resin in electron microscopy. It had a resin with relatively low viscosity (150 centipoise to 210 centipoise 25°C). Since Shell discontinued manufacture, replacements have been offered by EM supply houses that appear to behave similarly. Samples embedded in Epon or Eponlike resin formulations are easy to cut and routinely produce good ultrathin sections. As discussed later, the hardness of a sample block is strongly dependent on the ratio of anhydride to epoxy in the mixture and the type of anhydride present. Epon resins vary in this aspect, and the correct proportions can be determined from the epoxide equivalent (weight of Epon in grams that contains a 1-g equivalent of epoxy), which is usually listed on the label of the EM product. Since it is fairly hygroscopic, care should be exercised to keep Epon (and the sample) anhydrous throughout the preparation process.

Araldite, manufactured by Ciba Products Corporation, is another epoxy resin commonly used in EM sample preparation. It was one of the first epoxy resins to be used widely in TEM. Although the uncured resin is more viscous than Epon, sections from it are less grainy and more thermally stable. Araldite seems to have its cutting quality adversely affected by the inclusion of trace amounts of the nonreactive alcohols used in dehydration. This problem can be circumvented by using a compatible transitional solvent (such as propylene oxide) as the final step in dehydration.

A more recent addition to the choices for embedding media is vinyl cyclohexane dioxide (VCD). VCD (known commercially as ERL 4206) is an epoxy resin with the lowest viscosity (7.8 centipoise) of any used in EM. It is especially useful for material that is difficult to infiltrate (such as plant tissue with appreciable secondary cell wall development). It is readily miscible in the standard dehydrating agents. VCD is most often used in conjunction with another epoxy resin and additives in a series of mixtures that can produce blocks of widely varying hardness, depending on the amount of hardness modifier (DER 736) added. Sections cut from this material are very resistant to beam damage, but are difficult to stain with aqueous uranyl acetate solutions and usually require that the uranium stain be added either en bloc prior to sectioning or in an alcoholic solution. Once stained with uranium, the samples are easily stained with lead.

The second major component of a TEM resin-embedding mixture is the crosslinker or hardener, which is usually an organic acid dianhydride. Hardeners can serve two functions in the resin mixture. The first is to form covalent bonds between epoxy molecules, forming a highly insoluble copolymer matrix. The second function, in some cases, is to modify the hardness and therefore the cutting characteristics of the sample block. In Epon resin mixtures, for example, two anhydride hardeners—dodecenyl succinic anhydride (DDSA) and nadic methyl anhydride (NMA)—are added to the resin mixture in varying proportions to control the hardness of the final preparation. The longer nonreactive chain on the DDSA molecule is thought to impart more flexibility than the compact NMA molecule does; thus, mixtures relatively high in DDSA are softer. By varying the ratio of DDSA to NMA, an embedment can be made that is rigid enough to support a wide range of tissues without being unnecessarily hard.

Some resin mixtures rely on just one anhydride hardener. The VCD formula uses nonenyl succinic anhydride (NSA) as the hardener and other agents to modify the hardness. Similarly, most Araldite formulations use only DDSA as the hardener, with other compounds added to regulate block hardness. With any of the anhydrides, it is important to remember that they must be kept anhydrous to function properly in the resin mixture. Containers of complete resin mixtures that are stored in the refrigerator to retard polymerization should be warmed above the dew point prior to opening.

The third group of resin mixture components, those that modify the hardness of the sample block, can be subdivided based on how they react with the resin mixture. A plasticizer is a hardness modifying agent that is added to the resin mixture to decrease the hardness of the final block, presumably by filling space in the final matrix and decreasing the overall crosslinking. A plasticizer does not become a structural part of the copolymer system. Dibutyl phthalate, often used in Araldite formulations, is a plasticizer. As mentioned earlier, the ratio of two anhydrides used in the Epon resin mixture modifies the final block hardness.

Hardness modifiers that are structural components of the polymerized resin are called flexibilizers. In addition to using different anhydrides to achieve different hardness, varying amounts of different epoxies can be used to modify block hardness. In the case of the VCD formula, a small amount of the epoxy diglycidyl ether of polypropylene glycol (DER 736) is added to the mixture to soften the resulting block. This epoxy copolymerizes, with the VCD and NSA becoming part of the final matrix. The flexible aliphatic structure of this molecule is what reduces the hardness in the final polymer.

Finally, an accelerator is added to the resin mixture. Accelerators usually are substituted amines and do not actually form part of the polymeric matrix but catalyze the crosslinking reactions. Two common accelerators are S-1 (2-methyl aminoethanol) and DMP-30 (dimethyl amino methyl phenol).

Acrylic Resins and Modified Acrylic Resins The first group of resins applied to electron microscopy were the acrylic resins. Polymethyl methacrylate (known commercially as Plexiglas) is a hard, transparent material that sections with some difficulty. This problem can be solved by adding butyl methacrylate, which makes the block softer.

Although methacrylate mixtures have very low viscosity, infiltrate well, and can be made to section and stain well, they have two serious drawbacks for use in TEM sample preparation. First, upon polymerization they shrink considerably, causing components of the cells to pull apart severely. Second, methacrylate polymers lose mass (decompose) rapidly in the electron beam, which can lead to contamination problems in the microscope column. Also, as a methacrylate-embedded specimen thins down in the beam, it becomes distorted, further reducing the value of the remaining image. It is mainly for these reasons that the use of straight methacrylate embedding has been discontinued in most TEM facilities.

Although the plain methacrylates generally have been superseded by epoxy resins for routine work, one class of modified acrylics is becoming increasingly important for special applications. In particular, these compounds are useful in immunocytochemical reactions such as antibody gold labeling (discussed later in this chapter). Two such resin types are Lowicryl (K4M and HM 20) and the LR (London Resin

Company) (White and Gold) resins. These resins have the advantages of low viscosity, good solvent miscibility, and polymerization at low temperatures (with ultraviolet light). In addition, they appear to preserve the tertiary structure of proteins better and give lower background labeling than epoxy resins do. Ultrastructural details are often not as well resolved, however.

Once a suitable embedding medium has been chosen, the next step is to infuse the sample with the unpolymerized medium, which is most often done in a graded series generally similar to but with fewer steps than specimen dehydration. The length of time spent on each step depends on the sample, the resin, and the degree of agitation provided. Normally, the specimens are given from half an hour to overnight in each infiltration step. With any medium, especially the more viscous ones, adequate infiltration is greatly facilitated by some form of gentle agitation.

After the specimen has been transferred into 100% resin, it is usually beneficial to give it one to three additional changes of pure embedding medium to assure that the sample is completely infused and all traces of solvent are removed. Thus, the resulting sample block should be uniform in its cutting characteristics. The block is formed by embedding the specimen in a volume of fresh resin in a suitable mold with the specimen oriented so that the surface of interest is positioned conveniently at the edge of the hardened block. The casting is then hardened by accelerating the polymerization at elevated temperatures (e.g., 65°C) for epoxies or with ultraviolet radiation (e.g., 360 nm) for modified acrylics. Polymerization generally takes 8 to 72 hours for most media. The cutting quality of some resin formulations seems to improve with a day or so of postpolymerization aging.

Ultramicrotomy

After the specimen has been cast into a block of resin, the next step in sample preparation is to make ultrathin sections. For best resolution, the section of the sample viewed in the microscope should be as thin as possible. There are two underlying reasons for improved resolution with thinner specimens: First, the electron beam of the accelerating voltage used in most TEMs has limited penetration capabilities. Second, the depth of field of the objective lens in the TEM leads to the superimposition of detail of the specimen in the micrograph, making individual features hard to see with conventional micrographic means. Preparing good ultrathin sections is generally regarded as one of the most time-consuming and often most frustrating aspect of TEM sample preparation.

The finished ultrathin section should be less than 100 nm thick (ideally about half that thick) for most applications, and for best results the thickness should be as uniform as possible. Since a number of severe forces act on the section as it is being made, considerable care should be taken to avoid physical artifacts in the sectioning process. The final section should be free of compression, surface defects, chatter and knife marks, and dirt or precipitates acquired during the staining process. A good section should also be able to yield a dimensionally stable image long enough to take an electron micrograph.

Ultramicrotomes To obtain ultrathin sections, an ultramicrotome is required. These devices move the specimen block toward and over the edge of an exceedingly sharp knife in small, predictable increments. With currently available microtomes, advancing the specimen is accomplished in one of two ways (Figure 6.7).

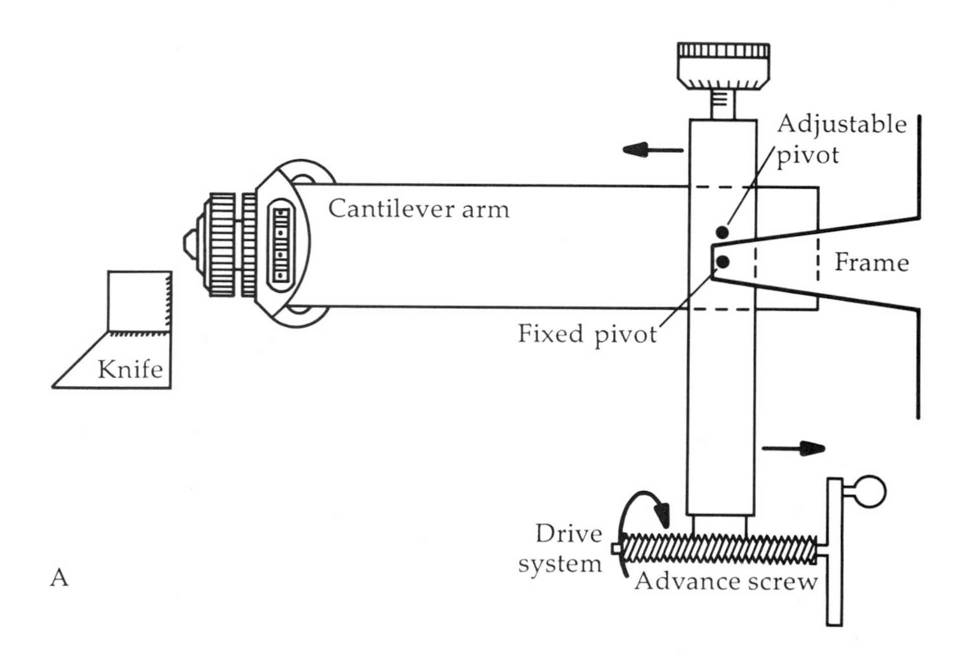

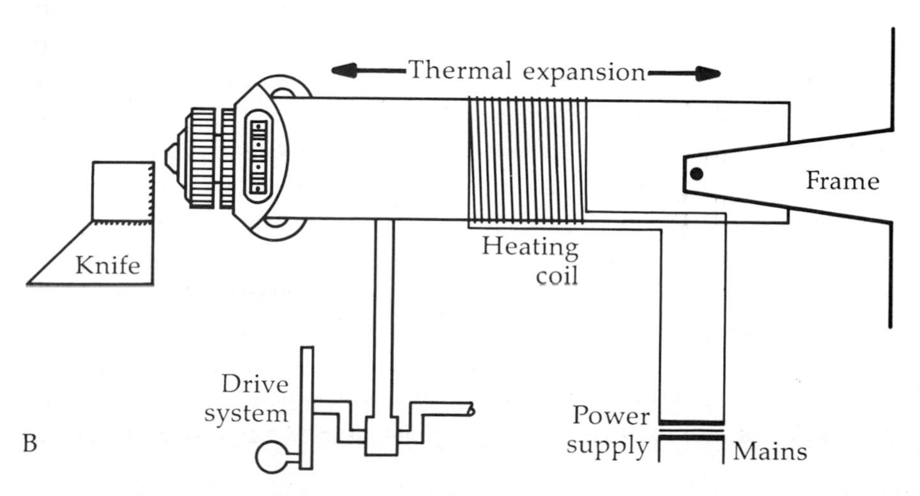

Figure 6.7 Ultramicrotome advance methods. (A) Schematic of thermal-advance microtome. (B) Schematic of a mechanical-advance microtome. Modern ultramicrotomes differ mainly in degree of automation; the basic mechanisms are the same. Because they are less temperature-dependent, mechanical advance systems are the best choice for cryogenic work.

The first method relies on the thermal coefficient of expansion of a metal. In this system, the metal arm on which the specimen is mounted is heated by a carefully controlled electric resistance heater. Since materials generally expand at a predictable rate over some temperature range, this thermal expansion can be translated into specimen advance. Since the heating is a continuous process, the periods between sequential cuts also must be controlled accurately. Both of these features are taken into account in modern thermal-advance microtomes.

Regardless of the type of thermal advance used, the expansion process cannot be protracted infinitely. The thermal-expansion properties of most metal alloys allow somewhat less than 1 mm of expansion for a 100°C temperature change in an arm of practical length for an ultramicrotome. Once this range has been exhausted, the arm must be cooled to reset the machine.

A second approach to specimen advance in an ultramicrotome is mechanical. In this type of microtome, the specimen arm is advanced through a series of levers, by the advance of a precisely machined nut on a similarly precisely made, fine-pitch lead screw. In this system, the amount of specimen advance is independent of the uniformity of the cutting cycle and depends only on the number of turns made by the mechanism. As with the thermal advance, there is a point at which the system must be reset, in this case involving the repositioning of the advance mechanism. Modern mechanical-advance ultramicrotomes are equipped with stepper motors and electronic controls that speed the sample through the cutting cycle, slowing to a precise cutting speed during a small period of the downward stroke.

With either type of microtome, the environment in which it is operated has a considerable influence on the ease and quality of sectioning. The microtome should be mounted on a stable, vibration-free surface (often supplied by the manufacturer) in a room free from rapid thermal fluctuations and drafts.

Ultramicrotome Knives The second major piece of equipment critical to making ultrathin sections is a knife with an extremely sharp edge. Although many steel alloys are hard enough to make knives for ultramicrotomy, none have been able to be worked to a sharp enough edge for ultrathin sectioning. At present, three materials are widely used for ultramicrotome knives.

The first material ever used (and still in use today) is glass. Polished plate window glass, cut into 1-inch squares, then broken diagonally, forms one of the sharpest edges known (Figure 6.8). The knife edge itself is formed by the natural cleavage of the glass and not initiated by a scored line. This is known as free breaking. Although this can be accomplished by careful use of an ordinary glass cutter and a pair of modified glazier's pliers, a more consistent approach is to use precut strips of plate glass in one of the commercially available knife breakers. Properly adjusted, these devices yield a high percentage of good-quality knives and are a worthwhile time saver in any EM laboratory.

Although glass knives are relatively inexpensive and are disposable, they do have some significant drawbacks. Even with modern knife-breaking machines, the quality of the knife edges produced varies considerably, especially of those produced by inexperienced workers. Second, the edge of the knife is very susceptible to damage by contact with any surface, which can occur easily, for example, when the section collection boat (a small trough filled with water on which the sections collect as they are cut) is being secured to the knife. The glass near the edge of the knife, being made of a supercooled liquid rather than a true crystalline solid, tends to flow, and the quality of the knife edge degrades with time. Finally, even when all goes well, the life of an area of the knife edge is limited. Usually about 20 to 30 sections of material can be cut before the edge becomes unacceptably dull.

An ultramicrotome knife can also be made from a piece of polished sapphire. These knives produce sections of good quality for a long time. They come mounted in a support that incorporates a boat, if they are set up for conventional TEM sectioning. The surface of the sections produced with these knives is smoother than that of a glass-sectioned specimen, which improves the overall resolution of ultrathin sections. Sapphire knives generally cost several hundred dollars, but do offer a considerable time savings over preparing glass knives. At present, no economic method of resharpening them exists.

For any significant amount of ultrathin sectioning, especially over a protracted period of time, a diamond knife is a worthwhile investment. Diamond knives consistently yield good ultrathin sections, and if used with care, will continue to produce good sections for years. The outstanding hardness of the diamond, combined with the ability to polish it to a fine edge, make this the optimum material for ultramicrotome knives (Figure 6.8).

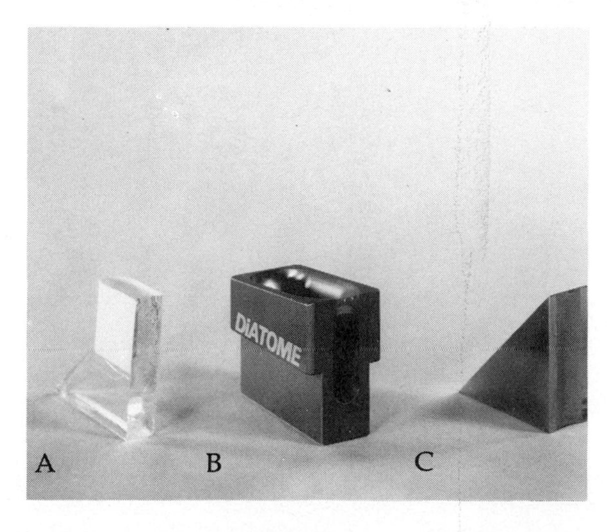

Figure 6.8 Ultramicrotome knives. (A) A glass knife suitable for both thick and ultrathin sectioning. (B) A diamond knife with a boat for conventional ultrathin sectioning. (C) A boatless diamond knife used for cryo-ultramicrotomy.

Of the knives mentioned, a diamond edge allows the user to set up the ultramicrotome and produce good sections in the shortest period of time. It also produces the smoothest section surface. Although the initial cost is the highest per millimeter of useful edge, the improved results and time saved are usually worth it. Although hard and sharp, the edge of a diamond knife is brittle and can be ruined by one mistake. Diamond knives without major edge damage can be resharpened at least several times, usually at half the cost of a new knife of similar length.

Although diamond and sapphire knives produce the best ultrathin sections, glass knives are still important in any laboratory doing sectioning. They are useful in rough trimming the specimen, in making thicker sections of the specimen for light microscopy, and in training new microtomists, prior to their using a diamond knife.

Steps in Ultrathin Sectioning The first step in ultrathin sectioning is to trim the specimen block down to a level where the specimen is exposed at the surface. This is normally done by holding the block in a microtome chuck under a low-magnification stereomicroscope and carving the resin away with a degreased, single-edged razor blade. If a few light-microscopy sections are to be made (a procedure that is useful in determining the quality of fixation and the area of the specimen for ultrathin sectioning), the specimen face should be trimmed up into a truncated pyramid with a face about 1 mm square (i.e., about the whole specimen). This block is then mounted in the microtome, and a few sections about 1 μm thick are cut, removed from the boat, and dried onto a glass slide for light microscopy (Figure 6.9).

Figure 6.9 Steps in preparing sections for light microscopy. (A) Specimen block trimmed for light-microscopy sections. (B) 1-μm sections cut with a glass knife (on the water surface). (C) Sections transferred to a glass slide. Evaluating thick sections to find the best area of the tissue for subsequent ultrathin sectioning saves time.

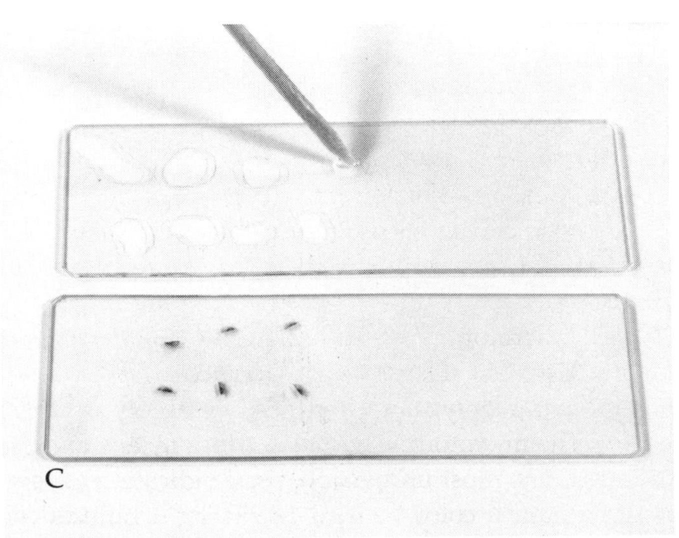

A

B

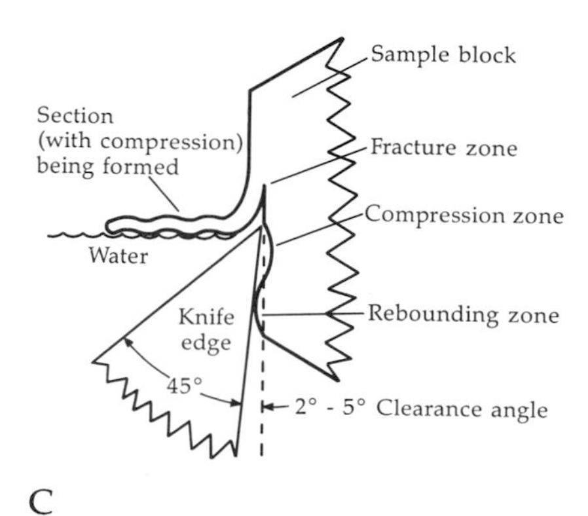

C

Figure 6.10 Steps in preparing ultrathin sections for TEM. (A) Specimen block retrimmed for ultrathin sectioning. (B) Ultrathin sections on the water surface. Thickness can be judged by interference color. A gold section is 80 nm to 100 nm thick. (C) The mechanics of ultrathin sectioning. (D) Retrieval of the sections with a grid.

D

Once the quality of fixation and area of interest have been determined, the block face should be reduced to a trapezoidal surface less than 0.5 mm on a side for successful ultrathin sectioning. The block is then remounted in the microtome, a fresh knife edge centered (or diamond or sapphire knife installed), and ultrathin sectioning begun.

If the block is good and the microtome has been set up properly, a ribbon of serial sections will start to accumulate at the edge of the knife floating on the water in the boat. This accumulation happens because as each section is cut from the block face, it persists at the knife edge until the next section is cut. Each new section adheres to the previous section and pushes it off the knife edge out onto the water (Figure 6.10).

All microtomes have a thickness control that can be set to a nominal desired thickness. Indications on the machine are generally a relative gauge only, however, and variability occurs from section to section. The microtomist must use other means to judge the specimen thickness. The most universally used indication of section thickness is the interference color formed as the light reflects off of the section

TABLE 6.2
Resin Section Thickness as Determined by Interference Colors

Light reflected from the water surface and the surface of the ultrathin sections have different path lengths. This leads to interference and the reenforcement of certain wavelengths, proportional to the section thickness.

Thickness	Color
≤ 60 nm	gray
60–90 nm	silver
90–150 nm	gold
150–190 nm	purple
190–240 nm	blue
240–280 nm	green
280–320 nm	yellow*

**At thicknesses greater than about 300 nm the sections first may appear red tinged but rapidly begin to look like a bit of cellophane floating on the water.*

surface and the water directly below it. This color gives an indication of the thickness to a tolerance of about 10% (Table 6.2).

Depending on the type of knife used and the consistency of the block being sectioned, the sections may arrive on the water surface with some vertical compression generated as the section is dragged over the knife edge. Since this feature would distort any measurements made on the final image and may be pronounced enough to cause visible ripples in the section, the compression should be removed. One method of removing sample section compression is to expose the surface to a small amount of solvent vapor by carefully holding a swab dipped in xylene over the completed sections. A second approach, and one that limits the microtomist's exposure to xylene, is to use a loop of heated wire (a heat pen). Whether chemical or thermal means are used, the sections are softened by the treatment, and the surface tension of the water acting on the hydrophobic section material pulls the compression out. Normally this removal of compression is accompanied by a visible shift of interference color to a thinner range.

Once the sections have been decompressed, they are picked up on either a film-coated or a plain grid. After the sections are retrieved, the grids should be placed in a clean, dry storage container until they can be processed further.

Specimen Staining Of Ultrathin Sections

Since biological material generally offers little electron contrast, sections are usually stained with an electron-dense stain (i.e., a reagent with a high degree of electron scattering, such as a heavy-metal salt) after they have been mounted on the grids. To work effectively, the stain must be able to penetrate the resin surface. If more than one stain is to be used, the stains must be compatible or synergistic with each other and differentially stain the various cellular components. The two most common EM stains for biological samples are based on lead and uranium and are most often supplied to the section as uranyl acetate and lead citrate.

Uranyl Acetate Uranyl acetate in a 1% aqueous or alcohol solution is used to stain sections first. The time it takes for proper staining depends mainly on the vehicle (alcohol is fast) and the resin used (VCD is slow). After the empirically derived staining time, the grid is given several washes in distilled water and carefully blotted dry. Uranyl acetate adds contrast to protein-rich structures and nucleic acids and provides some overall differential contrast to the rest of the cell components (Figure 6.11).

Lead Citrate Lead, usually in a form such as lead citrate, in a high-pH (about 12) solution is the most widely used stain for general EM investigations. It stains the glycogen, the SH groups of some proteins, nucleotides (via the phosphate groups), and compounds that react with osmium tetroxide (OsO_4). Lead staining is improved if the material has been stained previously with uranyl acetate because the

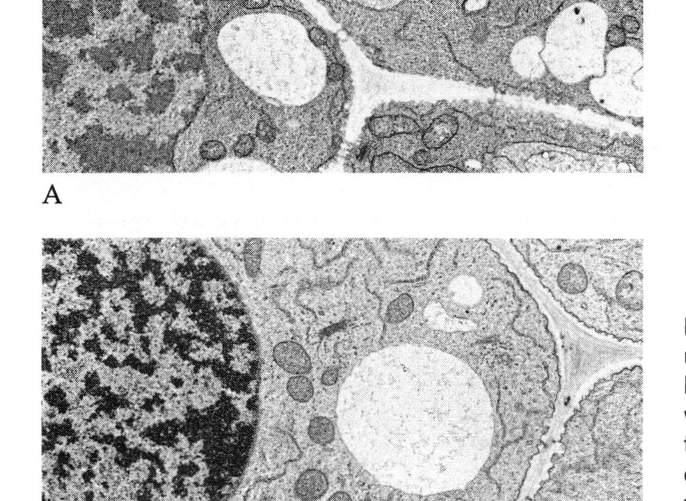

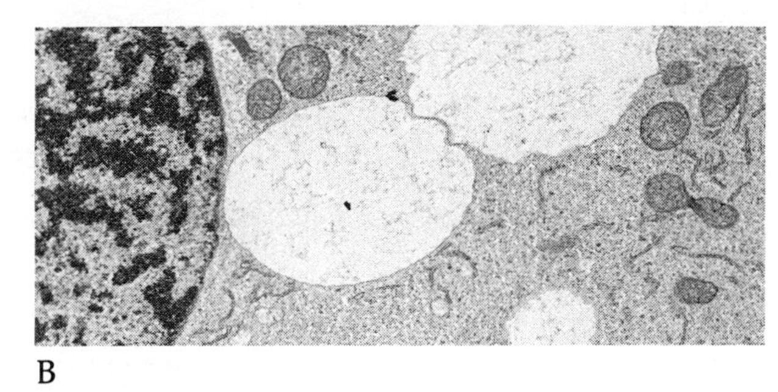

Figure 6.11 Effects of general stains on ultrathin sections. (A) An ultrathin section through an onion root cell, fixed with glutaraldehyde and OsO_4 and embedded in epoxy resin. (B) The same tissue with saturated, aqueous uranyl-acetate staining. (C) The same tissue with uranyl-acetate and lead-citrate staining. Note the sequential improvement in overall ultrastructural definition. Cell walls, especially, gain considerable contrast with lead staining after uranyl-acetate staining.

two stains appear to act synergistically. In plants, cell walls are left largely electron-lucent by pretreatment with both OsO_4 and uranyl acetate, but are stained darkly by lead citrate, although no known reaction with cellulose exists. Lead staining operations are prone to contamination by lead carbonate, formed by exposure to atmospheric CO_2. Contamination can be reduced by handling the staining operation in a chamber from which the CO_2 has been scrubbed by sodium hydroxide (NaOH) pellets. In contrast to uranyl acetate staining, lead staining generally can be accomplished in 1 to 5 minutes, regardless of the resin used. The stained image is defined better if the sections are washed briefly in a weak NaOH solution after staining is complete. This step also serves as a precipitation-inhibiting (high-pH) first rinse. After this rinse, the sections should be rinsed several times more to remove all traces of unreacted stain. Once dry, they can be viewed in the TEM directly or given a stabilizing coat of carbon (discussed later in this section) as an additional protective step.

Other Stains Although uranium and lead are the most common stains for TEM, some others are used for more specialized applications. Phosphotungstic acid, one of the first stains used, densely stains such components as mitochondrial matrices and muscle Z-bands. It also forms a complex with highly polymerized carbohydrates. (The reaction chemistry has not been worked out.) Ruthenium red, a stain also used in light microscopy, can be used in EM because of the electron-scattering power of the included ruthenium atom (atomic number 44). It stains pectic materials, especially after postfixation with OsO_4. It also reacts with acid mucopolysaccharides, some lipids, and probably some proteins.

In addition to electron-dense stains that react differentially but generally with cellular components, at present highly specific immunolocalization and enzymatic techniques (discussed later in this chapter) are being used more and more.

VACUUM EVAPORATORS AND EVAPORATION TECHNIQUES

One of the most important pieces of equipment in any EM laboratory is a vacuum coating device. This machine is also called a vacuum evaporator because evaporation is the method of forming the coating or thin film generally used. The vacuum evaporator is used for a number of sample preparation techniques in both SEM and TEM, including sample-support preparation, carbon-film coating, glow discharge, replica and particulate shadowing, and, when specially equipped, freeze-etch sample preparation. Many of the basic steps involved with the evaporator operation are the same no matter what the ultimate task is; it is thus expedient to discuss them first and together.

Several kinds of equipment can be considered vacuum evaporators (Figure 6.12). Often, when more than one is present in the laboratory, they have specialized jobs. The first type is the large bell-jar type. These free-standing machines have a glass bell jar supported on a metal base plate through which a number of vacuum-tight ports pass for the transmission of electrical power, and, in some cases, for cooling media and motion. They are usually pumped by a two-stage system using an oil-vapor-diffusion pump (see Chapter 3) to create the final vacuum. Once they are warmed up (about 30 minutes) and the sample has been inserted, these machines usually take from 5 to 15 minutes to pump down for a coating run at a vacuum of about 10^{-3} Pa.

A second approach to vacuum evaporation for the EM laboratory makes use of small, bench-top, high-speed pumping stations often built in a modular design with flanged work chambers of minimal volume. Interchangeable flanges are tailored to specific tasks. Since these evaporators are made with low-volume chambers and often are equipped with clean, high-vacuum pumps such as turbomolecular pumps, they can have a preparation ready to coat at a vacuum of 10^{-4} Pa in about 10 minutes, with no warm-up period.

The third type of evaporator used in EM labs is the freeze-etch machine. Freeze-etch machines range in construction from bell-jar evaporators that have been modified slightly for this task to dedicated small-chamber machines specifically for freeze-fracture and freeze-etch. Freeze-etch machines are available with oil diffusion-pumped, turbomolecular-pumped, and cryopumped systems and with an array of preparation configurations and options.

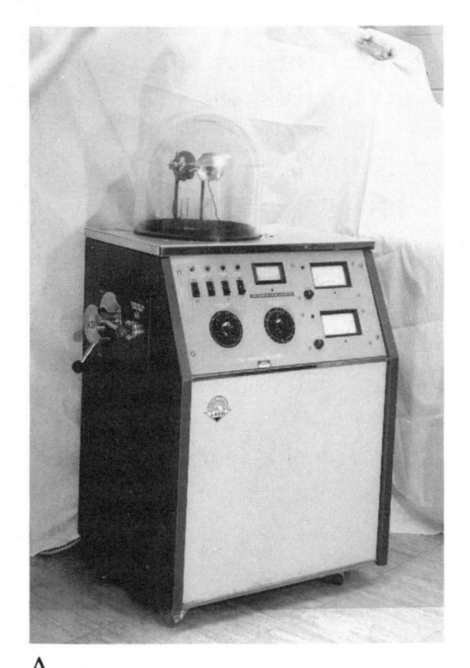

A

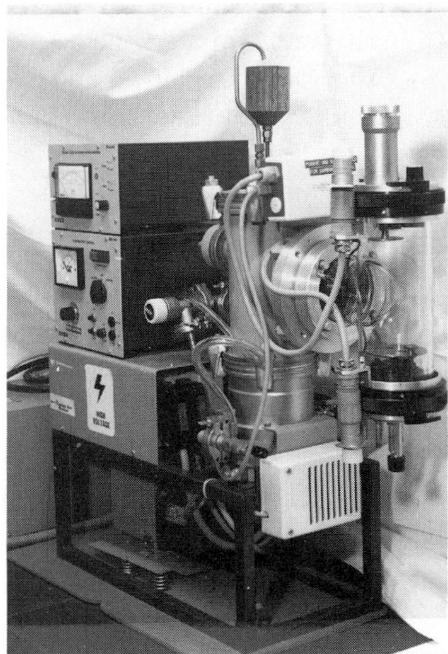

B

C

Figure 6.12 Common vacuum evaporators. (A) Bell-jar type. (B) Modular bench-top type. (C) Early freeze-etch unit. All of these machines can be used for routine EM laboratory evaporation jobs. The freeze-etch machine has provisions for sample fracture, etching, rotation, and precisely controlled shadowing at temperatures from −196°C to room temperature.

Requirements for Vacuum Evaporation

Regardless of the type of evaporator unit or the task at hand, the preparation technique has some general requirements. As described earlier, the first requirement is a source of vacuum to form thin films of solid materials by evaporating them from a source and depositing the vapor as a layer on the preparation. In some cases, the vacuum is also required to reduce contamination of the specimen surface during the coating operations.

Within the vacuum chamber, a mechanism to accomplish the evaporation process is required that can convert a usually stable solid with high melting point into a vapor in a controllable way. The material to receive the thin film application must itself be stable in a vacuum. Thus, it must be free of trapped, included, or adsorbed gases or substances with low vapor pressures. In some applications, however, these substances may be tolerated if their influence can be diminished by temperature regulation. For example, if water is introduced into a vacuum-evaporator system, it takes considerable time for the pumping system to draw it off in a vapor. If the water were frozen and maintained at a very low temperature, however, the high vacuum could be achieved more rapidly.

Thermal regulation is important with some samples. Although no appreciable amount of heat can reach (or leave) the sample by convection (the vacuum present is much greater than that of the best thermos), some heat flux is possible by conduction through the support. Radiant heating, especially during the evaporation process, is usually the major factor. Cooling a sample or maintaining a low temperature can be accomplished by using a temperature-controlled specimen-support stage in the vacuum chamber.

Finally, for many applications, it is important to know the thickness of the film deposited on the sample. Determining this thickness can be as simple as observing a color change on a known substrate or as complex as employing elaborate electronic monitoring devices that can be integrated into the evaporation system.

Methods of Evaporation

The oldest method of evaporation in the vacuum evaporator is by electrical resistance heating. In the resistance-heated evaporator, a single filament, wire basket, or thin trough of a metal with a high melting point (usually tungsten) is wrapped or filled with the substance (usually a metal with a lower melting point) to be evaporated (Figure 6.13A). The filament assembly is placed between two electrical contacts at some predetermined geometry from the sample, and a current is run through

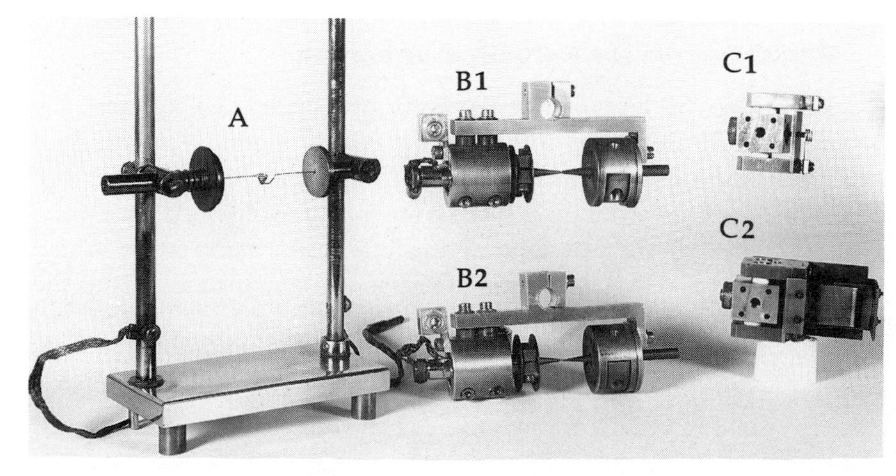

Figure 6.13 Evaporation sources for preparing thin films. (A) Tungsten-wire resistance-type sources are inexpensive but are harder to control and must be reassembled after each application. (B) Carbon-arc sources for (1) plain carbon and (2) platinum/carbon evaporation. These are better than resistance methods, especially for platinum evaporation. (C) Electron-beam gun. (D) Electron-beam gun (top removed), showing electron-emitting coil and platinum/carbon electrode for evaporation. Electron-beam evaporation provides the best combination of reliability, thermal protection, efficiency, and film quality.

it. As the current flows, the support wire heats up, and the supported metal evaporates. The method can be adapted to carbon evaporation by using a braided carbon cord instead of the metal filament.

This method of evaporation is the simplest and, in general, also the cheapest. It can be used to deposit thin films of gold, gold/palladium, platinum/palladium, and, by heating the filament alone, tungsten. The major drawbacks to resistance evaporation are that the evaporation device usually needs to be reassembled for each run, the process of evaporation produces a high heat load on the specimen from the incandescent filament, and reproducing the film thickness accurately from one run to the next is difficult. Another problem is in evaporating platinum from a tungsten wire. Since the melting points of the two metals are so close to each other, the platinum often melts into a drop that alloys with the tungsten wire. The wire then burns through at hot spots formed on either side of the drop before much of the platinum can be evaporated.

A second method of evaporating materials for EM sample preparation is to use a carbon-arc-type device (Figure 6.13B). These evaporators work by applying a low-voltage, high-amperage current through the faces of two carbon rods, which are sprung in end-to-end contact. The application of current causes vaporization of the carbon at the interface and the formation of a high-temperature electrical arc that vaporizes the carbon substrate, which is then deposited on the target.

To form a metal film, a coil of the desired metal (usually platinum or a platinum alloy) is wrapped around the end of one of the carbon electrodes and is rapidly evaporated by the heat of the arc. Films so produced are mixtures of carbon and metal, but if properly done, the carbon content is usually low enough to be operationally insignificant. This type of evaporation is significantly faster for depositing platinum films than direct resistance heating is and therefore reduces the heat load on the specimen considerably. With careful operation, more than one run can be made before the apparatus needs to be reset. This is especially true for plain carbon evaporation. The evaporators themselves are relatively inexpensive, but as with the resistance methods of evaporation, the precise control of film thickness is difficult.

A third type of evaporation device is the electron-beam gun (Figure 6.13C). These evaporators work on similar principles to the thermionic emitter of a tungsten-filament EM electron gun. Electrons emitted from a coiled incandescent filament are accelerated toward the material to be evaporated by a potential of several kilovolts. On collision, the kinetic energy loss is converted to heat, vaporizing the end of the anode, which is made of the film substrate. The evaporator-gun body is often fitted with a set of grounded ion collectors that extract charged particles from the vapor stream. Electron-beam evaporators work well for generating thin films of platinum, tungsten/tantalum, and carbon. As with the carbon-arc devices, the platinum films made with electron-beam guns contain some carbon, usually about 5%.

Electron-beam guns have some distinct advantages over the resistance and arc evaporation methods. Typically they can be used for six or more runs before requiring resetting or reloading. Once the acceleration voltage and emission current are optimized for a given application, the deposition rate is quite stable, and uniform thin films can be applied by timing the gun burn. Since the heating is localized to the source tip within the gun assembly, the thermal load on the specimen is lower than with the other types of evaporators. As with the diamond knife used in thin sectioning, the higher initial cost of the electron-beam evaporator equipment should be offset easily by the increased productivity in almost any application.

Specimen Requirements for Vacuum Evaporation

Any sample to be processed in the vacuum evaporator must have certain characteristics for the process to work. First, the sample must be stable in a vacuum. It should have as little trapped gas as possible to prevent problems with dimensional stability, as well as to avoid destruction of the vacuum as it outgases, which would greatly increase the time to pump down the vacuum chamber to a working vacuum. A major source of residual gas in the specimen preparation evaporator is water vapor. Samples and evaporation equipment should be kept as dry as possible. Another contaminant to watch out for is oil with a low boiling point (such as skin oil), which must be pumped from the system. In addition to increasing the pump-down time, the presence of condensable vapors in the bell jar can lead to specimen contamination. This is especially important in applications where the specimen is held at low temperatures, such as freeze-fracture, because the cold specimen surface (which is to be replicated) acts as a trap for these residual vapors. Using cloth gloves to handle materials bound for the evaporator helps reduce these oils.

The materials used for preparing support films (such as Formvar and Parlodion) generally pose no problems and can be coated with carbon easily in the vacuum evaporator. After staining, thin sections in epoxy resins can also be coated lightly with carbon in the vacuum evaporator to enhance beam stability. Critical-point-dried material (see Chapter 7) for SEM or TEM generally has passed through sufficient fixation and solvent steps that it too can be processed in the vacuum evaporator. Some material, such as molecular samples that are applied to their substrate prior to shadowcasting (discussed later in this chapter), can be introduced directly into the vacuum, since the total water content is insignificant.

Film Thickness Must Be Known for Critical EM Work

Just as judging the thickness of the section is an important aspect of ultramicrotomy, so is measuring the thickness of the film an important aspect of evaporator techniques. Film thickness determines the strength and quality of carbon films and the resolution and contrast of shadowcast samples. A number of methods for measuring film thickness in TEM sample preparation exist.

The first and easiest method is simple observation. Metal or carbon thin films impart a visible change in color to the specimen or target. A small piece of white paper included with the samples provides a suitable surface to measure the thickness of the films optically. An experienced observer can judge not only the thickness of the film, but also its composition (in the case of platinum/carbon evaporation) by the color of the evaporated film.

A second approach to determining the film thickness is quantitative evaporation. If the distance from the evaporation source to the specimen, the density of the metal being evaporated, and the mass of the metal used are known, the thickness of the film theoretically can be calculated based on the idea that the evaporated material deposits itself as though on a spherical surface with the source at the center. Since the evaporator filament retains some of the metal and other inefficiencies may exist, this system rarely agrees with the measured film thickness. Once the adequacy of a certain amount of metal at a given set of evaporation variables is determined, however, the reproducibility is usually good.

Quantitative measurement of the film thickness during application can be accomplished in at least two ways. The first method makes use of the conductive nature of the evaporated films. A nonconductive surface with two electrodes at a fixed separation with a predetermined potential is inserted into the evaporator path. As the film thickness increases during the evaporation process, the impedance diminishes, and a current proportional to the film thickness flows. This current can be monitored remotely.

Although this technique works, the short useful life of the detector is a major shortcoming. Since the films are highly conductive, resistance quickly falls as the thickness of the film increases and the sensitivity (change in resistance) decreases. A second problem is that the thickness for each resistance value must be derived empirically. The system is, however, simple and inexpensive to build and can be incorporated into the evaporation control circuit.

A second indirect approach to measuring film thickness uses the change in the electrically induced oscillation of a quartz crystal to monitor the amount and rate of film deposition. The quartz-crystal monitor measures film thickness by comparing the frequency of a reference crystal to that of a similar crystal exposed to the evaporated film. As the film thickness increases, the resonant frequency of the crystal drops. This change is uniform over a considerable range, and the monitor is sensitive enough to detect a difference as small as a single atomic layer. Commercial models of quartz-crystal monitors are available from a number of manufacturers, and when coupled as a controller to electron-beam guns, provide a reliable way to produce uniform film coatings consistently.

SHADOWCASTING AND REPLICA TECHNIQUES

Metal shadowcasting is one of the oldest, and still very useful, TEM sample preparation techniques. It uses the vacuum-evaporation equipment described earlier to contrast ultrastructural topographical details in electron-lucent biological material by applying a nonuniform coating of a heavy metal. The thickness of the metal is greater on the side of a positive topographical feature toward the evaporator source; thus, it scatters more electrons and appears darker than the other side, which is shielded from the evaporator source. Of course, this technique requires the specimen to have some surface features of interest. Shadowcasting is ideally suited to small biological materials, such as macromolecules and viruses, and is a fairly rapid preparation technique that yields an exceptionally stable sample. If the specimen is too large or electron-dense to shadowcast directly, an electron-lucent replica often can be made that can yield information about the exposed surface.

Shadowcasting

To shadowcast a specimen, the microscopist must first determine its suitability. The specimen must meet the requirement of stability in a vacuum mentioned earlier. It must be electron-lucent, as the specimens in the case of negative staining (discussed earlier) also must be. Of course, the specimen needs to have surface details or morphology of interest. Finally, the specimen must be provided with a support film.

Samples may be prepared for shadowcasting in a variety of ways. If the sample is a dry powder, it can be simply dusted onto a support film. (Plastic films seem to work better than carbon or carbon-coated films.) If hydrated, the material may be applied from an aqueous solvent, dropwise, onto a support film in a method similar to that already described for negative staining. After the residual solvent dries, the specimen can be introduced into the vacuum evaporator (or the evaporator can be used to dry the sample quickly). Once a working vacuum is established, the sample is shadowed using one of the evaporation techniques mentioned earlier. This method is simple but has potential shortcomings. If the specimen requires a buffer to maintain the desired morphology, the buffer must be volatile (e.g., ammonium/formate or pyridine/acetate) or the sample is likely to be obscured by or decorated with crystals of the buffer salts. Any contaminating material in the sample solution also will obscure details. Finally, if the specimen is long and highly flexuous, as with many biomolecules, it may not remain in a useful configuration as it dries but instead may dry in a nondescript mass.

One way to improve the probability that the details of a highly extended macromolecule are visible is to use a molecular spreading technique. A common example is the Kleinschmidt technique for spreading the highly flexuous and long DNA molecules. In this technique, the DNA (in the volatile buffer ammonium acetate) is mixed with a protein (cytochrome *c*), and the mixture is gently applied to a clean water surface. The denatured cytochrome *c* forms a film at the air/water interface to which the whole length of DNA is attached. The DNA is thus oriented in a single plane and can be picked up on a coated grid and shadowed to show its length and other details such as nicks, loops, and number of strands. Caution should be used in determining the size of the molecule, since it is covered with the cytochrome *c*.

Another application technique is to spray the specimen solution onto the support film with a nebulizer or modified airbrush. This technique allows for a more precise application rate and can overcome some of the adsorption problems that exist with direct application to thin films. In this technique, the specimen in solution (in solution) is applied in the form of a mist of small droplets to the support surface and then placed in the evaporator for shadowing. A variant of this technique uses a glycerol solution to carry the specimen, which is sprayed onto a freshly exposed (i.e., cleaved) mica surface. The molecules are blotted against the mica as the spray droplets hit the surface, and a remarkably clean preparation results. In this case, the whole mica chip is shadowed, and the support film is formed by applying a continuous layer of carbon over the discontinuous shadowing film. This carbon-backed preparation can be scored into squares, floated off onto water, and mounted on grids as was described earlier for the pure carbon-film preparation technique.

Once the specimen is mounted on the coated grid or substrate in the evaporator and a suitable working vacuum has been reached (ideally less than 5×10^{-4} Pa), shadowing can proceed. A number of metals can be used for TEM shadowcasting (Table 6.3); those that have proven to be most useful are platinum, chromium, gold, tungsten, and alloys of tungsten/tantalum or platinum/palladium.

There are several shadowing geometries. The first and simplest is to shadow the specimen from an angle of 45° from a point source using one of the evaporators mentioned earlier. Here the metal accumulates on the side toward the evaporation source, creating areas of higher electron density corresponding to height differences and shadows (light areas) on surfaces away from the evaporator source. The 45° angle not

TABLE 6.3

Materials Commonly Used in Shadowcasting and Coating for TEM and Their Boiling and Evaporation Temperatures

In addition to lowering the temperature required for evaporation, the vacuum environment of the evaporator provides a long mean free path for the evaporated material being deposited and keeps the vaporized materials from reacting chemically with air gases.

METAL	BOILING POINT (°C at 10^5 Pa)	APPROXIMATE EVAPORATION TEMPERATURE (°C at 10^{-2} Pa)
Platinum (Pt)	4530	2020
Tungsten (W)	5900	3030
Gold (Au)	2700	1405
Carbon (C)	4827	3367
Tungsten/Tantalum (W/Ta)	≈ 5500	3030
Chromium (Cr)	2672	1800

only makes an image that is easy to interpret but also makes contour approximations easier, since the height is nearly equal to the shadow length (Figure 6.14). For specimens with very little vertical relief, it is often advantageous to use a smaller angle. For macromolecular studies, for example, angles as small as 2° to 5° are often used.

In general, the average thickness of the contrasting metal film should be 1 nm to 2 nm to allow optimum contrast and resolution, although higher resolution may be achieved with thinner coatings with some metals. Thus, the thickness of the film is an important considera-

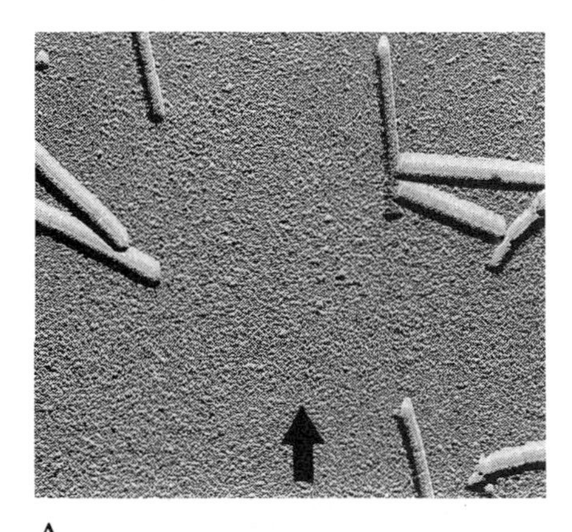

A

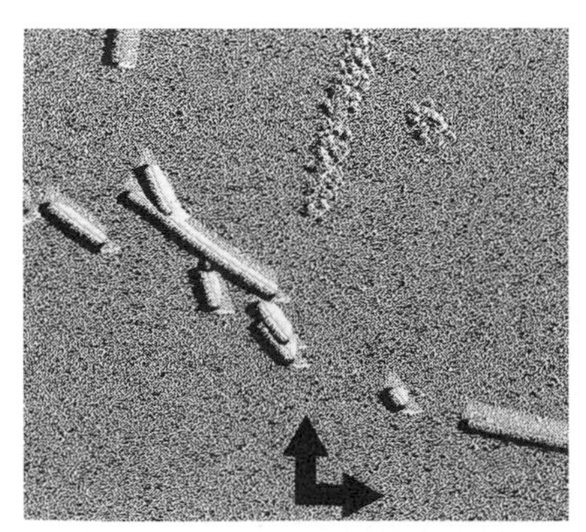

B

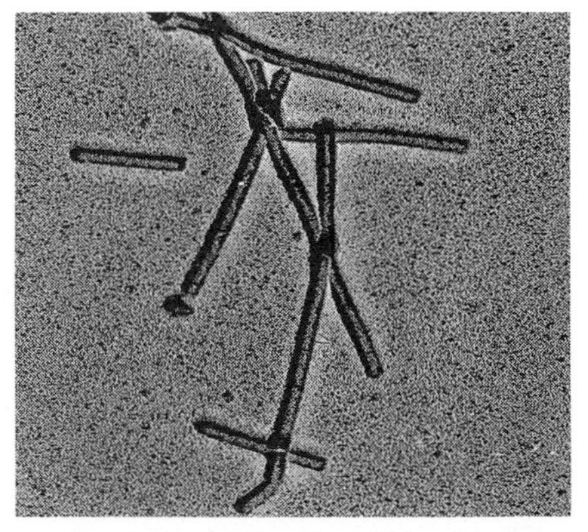

C

Figure 6.14 Effects of shadowing geometry. (A) Tobacco mosaic virus (TMV) unidirectionally shadowed at 45° with platinum. (B) TMV portrait-shadowed. (C) TMV rotary-shadowed. Portrait and rotary shadowing reveal more of the specimen surface than unidirectional shadowing does but produce shadows that are less distinct.

tion in making measurements at the limit of resolution for shadowed preparations (which is about 2 nm to 3 nm for most materials). Using two metal evaporation sources placed 90° to each other on the vertical axis of the specimen can produce portrait shadowing, which, with a unidirectional approach, reveals details hidden in shadow. This approach can be quite informative with samples having many angled surfaces (such as icosahedral viruses) or long lengths (such as DNA).

Another variation on the shadowing technique involves rotating the specimen during the shadowing process. Rotation makes it easy to determine the continuity of fibrous samples and to detect details not visible with the unidirectional approach. If very small angles are used, rotary shadowing allows the investigator to build up the height and width of molecules while adding little background contrast (see Figure 6.14).

If the specimen has been mounted on a grid covered with a support film prior to shadowcasting, it is ready for viewing in the TEM on completion of shadowing. If the specimen has been shadowed on some other substrate, such as mica, it can have a carbon backing (support film) applied while in the evaporator by evaporating carbon from a source directly over the specimen.

Replica Techniques

Many TEM samples with surface details of interest are too thick (more than 100 nm to 200 nm) to allow the use of the shadowcasting techniques described in the previous section. In many cases the desired information can still be extracted by making a surface replica, which can be accomplished in several ways. Replicas may be made either directly or indirectly, depending on the characteristics of the specimen and the information desired. For most biological applications and some material science applications, the single-stage or direct replica method is used.

Single-stage replicas are made directly from the surface of the biological sample. The sample material, suitably dried, is introduced into the evaporator with the surface of interest facing up. If the specimen is too small to handle directly, it can be mounted on a piece of mica first. After a vacuum in the $\geq 10^{-4}$ Pa range is reached, the specimen is shadowed as described for shadowcasting, usually with platinum. Immediately after the metal shadow has been applied, the whole specimen is coated with a layer of carbon, usually about 5 nm to 20 nm thick. This backing supports the discontinuous metal film in contact with the specimen.

The specimen with the replica attached to its upper surface is floated off onto a water surface and then transferred to a cleaning bath. The type of cleaning bath depends on the nature of the specimen material and the type of shadowing metal. Platinum/carbon replicas can withstand almost any chemical cleaner. Common choices are liquid bleach (sodium hypochlorite solution), hydrogen peroxide, and chromic acid solutions. For specimens that have a significant mineral component or that have been mounted on glass, hydrofluoric acid (HF) can be used to liberate the replica. Care should be used in cleaning delicate replicas; the cleaning fluid may need to be diluted to keep agitation caused by the chemical reactions to an acceptable level. After being cleaned, the replica is rinsed several times in distilled water and mounted on a copper grid (either coated or uncoated) and is then ready for viewing in the TEM. The single-stage replica technique works well for cell surfaces with a single layer, such as leaf surfaces or cultured cells grown on a flat substrate, such as a coverslip.

Indirect replica techniques are used less frequently than are direct methods in the biological sciences. They can be of utility if the original specimen cannot be destroyed, is too large for the vacuum chamber, or is not stable under the vacuum conditions. Usually a 3% to 4% solution of cellulose nitrate (e.g., Collodion) in amyl acetate is used to form the first replica stage. This solution is applied to the original sample, allowed to dry, and then stripped off. The surface can either be replicated with carbon that is shadowed, or it can be shadowed directly and then backed with carbon to form the final replica for TEM observation. This final replica is cleaned of the plastic with an appropriate solvent.

Freeze-Fracture and Freeze-Etch Techniques

A final variation of the shadowcasting/replica techniques is freeze-fracture. This process involves splitting a cryogenically stabilized specimen and subsequently shadowing the freshly exposed surface. Freeze-etch is a slight modification of this technique in which a small amount of the ice at the freshly fractured surface is sublimed away to reveal details hidden below the fracture plane. Freeze-fracture allows the observation of intracellular details in three dimensions and has the unique property of splitting through the middle of membranes encountered in the fracture plane of biological tissue samples. Thus, it is especially useful in characterizing and localizing membrane-bounded macromolecular structures. In addition, since the replica is made directly from frozen hydrated material, studies of materials unaffected by chemical fixation can be made if the appropriate ultrarapid freezing techniques are used. Although the process and the interpretation of the resulting images requires at least a cursory understanding of the underlying properties of vacuum and thin-film generation, freezing processes, and the behavior of biomolecules at cryogenic temperatures, freeze-fracture provides information not accessible by other means.

Requirements for Freeze-Fracture As with other replica techniques, the sample preparation equipment must meet certain requirements for freeze-fracture and freeze-etch. These requirements are more stringent for freeze-etch than for most other replica techniques.

Since the sample to be replicated is hydrated, it must be stabilized by freezing, both to fix the specimen and to allow its introduction into a high-vacuum environment. One of the ultrarapid freezing methods described earlier should be used so that the ice crystals are below the resolving limit of the contrasting metal film.

A second important consideration is the level and quality of the vacuum in the vacuum system. Since the specimen is at temperatures below –100°C, it can form a cold trap and collect oil and water vapors present in the residual gases. Vacuums need to be at least in the low 10^{-4} Pa range.

The sample must fracture along a plane of interest. With most biological samples, the material fractures through the middle (the hydrophobic region) of the membrane. The specimen should fracture with a minimum of plastic deformation, which is often related to the specimen temperature. Sometimes fracturing at a low temperature and then warming the specimen slightly before coating it gives better results.

If information on membrane surfaces (or other hydrated surfaces) is desired, the specimen must be etchable (i.e., the frozen aqueous phase must be able to sublime to reveal details below the fracture plane).

Since the sample is so cold and vulnerable to condensation contamination, the evaporation technique used must rapidly deposit the coating films without significantly heating the fractured surface of the sample. To avoid overheating and, more important, to prevent the obscuring of valuable detail, a means for judging film thickness accurately is desirable. Electron-beam evaporators for depositing the film and a quartz monitor for judging its thickness are highly advantageous.

Finally, it must be possible to clean the replica. Cleaning materials must be available that are sufficiently powerful to remove the residual specimen from the replica without ruining it.

Freeze-Fracture Procedure The freeze-fracture process consists of six essential steps (Figure 6.15). The steps, similar to those previously described for TEM specimen fixation and replica techniques, are described here only briefly, but their importance should not be overlooked. The first step of freeze-fracture is to isolate a representative sample from the specimen. Since the ultimate size of the sample will be quite small, the cautions on prefixation mentioned earlier are even more important. One of the most common problems with sample preparation is specimen dehydration occurring between the isolation and freezing steps. The reduction of water in the sample can lead to results that, at first, appear better than they really are because the formation of ice

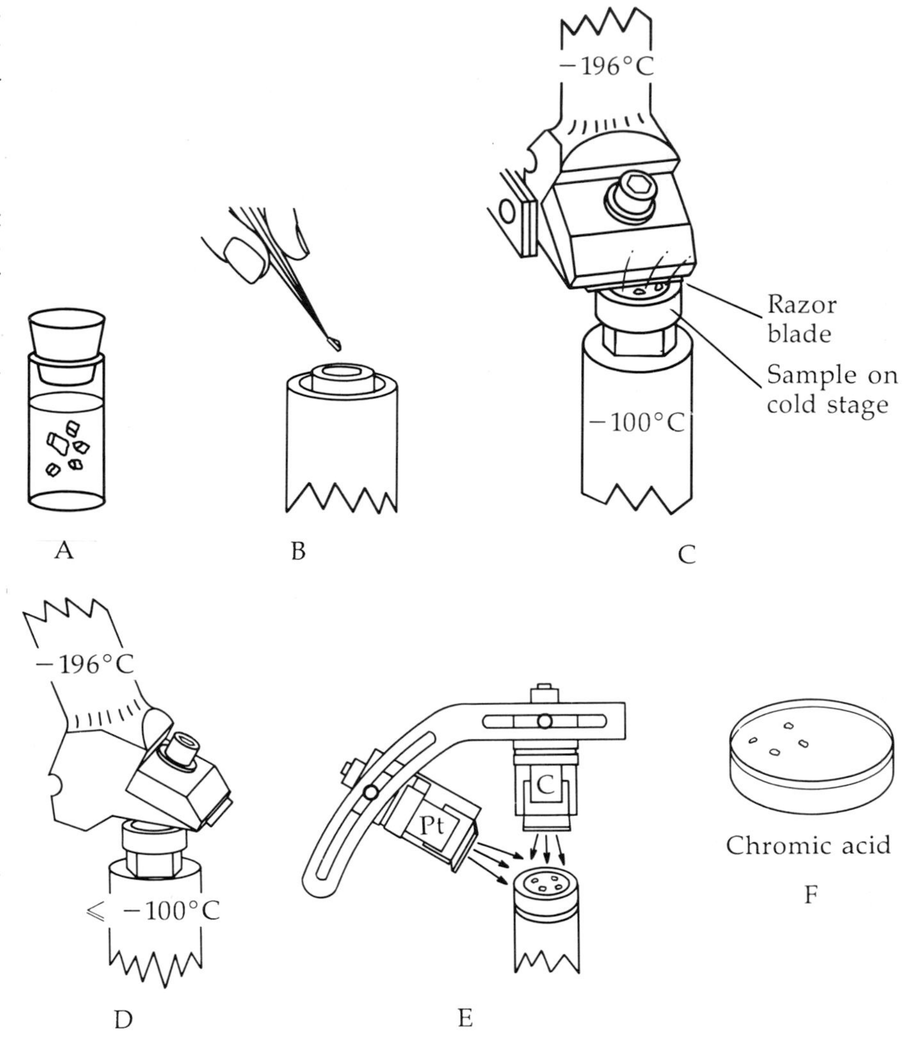

Figure 6.15 Six essential steps in freeze-etch. (A) Specimen isolation and pretreatment. The quality of the final replica hinges on this crucial step. (B) Ultrarapid freezing. If the sample has been cryoprotected, plunge freezing works well with sample sizes of nearly 1 mm^3. (C) Fracturing. This step can be done with a knife at −196°C or by tensile stress with the snap trap. (D) Etching (sublimation of water from the fracture surface using the microtome knife as a cold trap). (E) Shadowcasting and carbon-backing. (F) Replica cleaning. All biological material must be removed to produce a clear image in the TEM.

crystals is suppressed at the higher solute concentrations in the drying tissue. It is essential to have the whole isolation process worked out step by step before any samples to be replicated are frozen.

There are several methods of ultrarapid freezing once the sample has been isolated. Selecting a method is usually based on suitability to the specimen and availability of equipment. All of the freezing techniques that operate at atmospheric pressure have very limited sample thickness capacities. To overcome this limitation, a cryoprotectant is sometimes used. (Often this is glycerol, which is serially infiltrated into the specimen to a final concentration of about 30%.) To avoid gross disruption of the ultrastructure, a light glutaraldehyde fixation treatment (e.g., 0.5% to 2.5% for 30 minutes) is generally carried out prior to glycerol infiltration. If the sample is to be etched, the use of glycerol is precluded, since glycerol solutions etch very slowly.

After the sample has been isolated and pretreated, it is mounted on a carrier for the freezing process. The type of carrier used is dictated by the freezing apparatus and the freeze-fracture equipment. For plunge freezing, small gold disks with a raised specimen retaining wall are usually used. The specimen is placed in the center, and the assembly is plunged into the cryogen as described earlier. Other supports, such as a specimen support made of thin (approximately 80 μm) copper, may be used for this process. The thermal conductivity of pure copper is excellent.

Similar thin copper supports are used for propane-jet freezers, but with propane-jet freezers the specimen (which must be less than about 40 μm thick to achieve the optimum effect) is sandwiched between an upper and lower support. Supports for the metal-block freezer are generally about 10 mm in diameter, are made of copper or aluminum, and often are designed to fit custom-tailored freeze-etch stages.

Once the sample has been frozen ultrarapidly, it is usually stored under liquid nitrogen, until it can be loaded into the freeze-fracture machine. The time between freezing and storage under liquid nitrogen should be as brief as possible (no more than a second or so) to prevent recrystallization.

After inserting the specimen into the freeze-fracture machine, the fracturing process can be accomplished in two ways. The classical approach is to shave thin layers off the top of the specimen with a microtomelike knife cooled to liquid nitrogen temperatures (Figure 6.16).

This method usually works well, but some samples are brittle enough that one pass causes numerous fractures in the specimen, which can become contaminated with water vapor. Instead of shaving off thin layers of the specimen, one large slice can be cut, thereby reducing this type of prefracture artifact. The best technique is usually arrived at

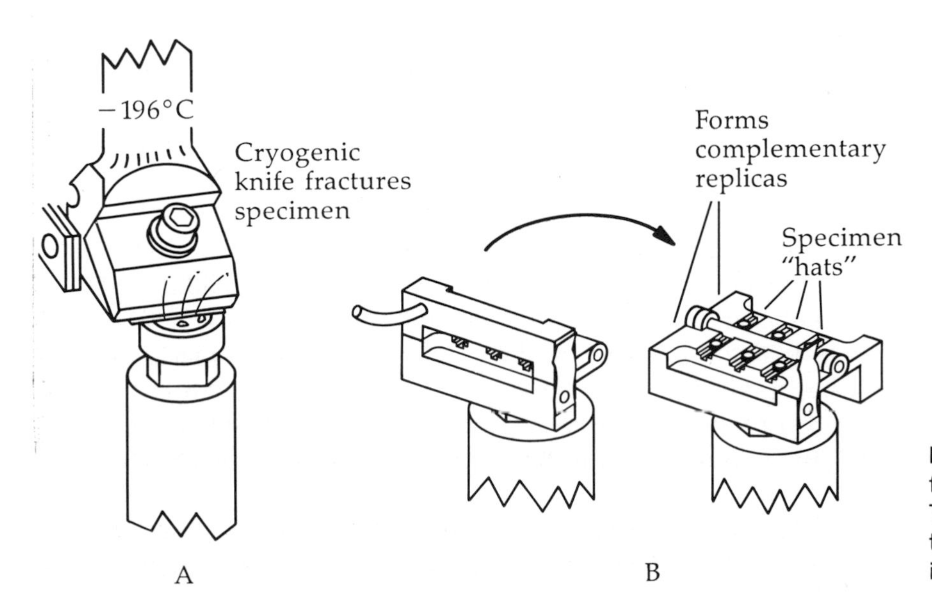

Figure 6.16 Fracturing methods for freeze-fracture. (A) Cold microtomy. (B) Tensile stress fracturing. Two surfaces that can be replicated are formed, allowing complementary replicas to be made.

TABLE 6.4
Sublimation of Pure Ice in a Vacuum
Specimen temperature control is crucial during freeze-etch. A 10°C change in temperature can lead to a tenfold change in etching rate.

TEMPERATURE (°C)	RATE (nm/s)	VAPOR PRESSURE OF ICE (Pa)
– 90	14.0	6.7×10^{-5}
–100	2.3	1.0×10^{-5}
–110	0.26	1.2×10^{-6}
–120	0.02	9.4×10^{-8}
–130	0.001	5.6×10^{-9}

Data from Koehler, J. K. 1968. The techniques and applications of freeze-etching in ultrastructure research. In Lawrence, J., and Gofman, J.W., eds. Advances in Biological Medical Physics, Vol. 12, *pp. 1–84. Academic Press, New York.*

empirically. If the specimen was frozen by a method in which it is sandwiched between two supports, the fracturing process can be done by tensile stress-fracturing: The sample is simply pulled apart. Many sample-stage designs allow this practice. The two replica surfaces produced are complementary. Analyzing complementary figures gives information not only about intramembranous particles, but also about some of the subtle artifacts that can be generated.

After the specimen is fractured, the exposed surfaces can be shadowed immediately, or they can be etched lightly, removing only a few nanometers of ice off the surface by sublimation.

Etching is done by heating the specimen stage to a temperature above that of liquid nitrogen (–196°C) but below one that would promote recrystallization of the ice (about –80°C). The rate of sublimation is also vacuum-dependent: The higher the vacuum the greater the sublimation rate (Table 6.4). These two environmental conditions are balanced to yield an acceptable rate, usually about 2 nm/s. To drive the sublimation and keep water from recondensing on the specimen surface, the cold microtome knife is usually positioned over the specimen for the etching period (1 to 3 minutes).

Following fracture and the optional etching step, the surface of the sample is shadowcast with a heavy metal (usually platinum) by an electron-beam evaporator. The vacuum for this operation should be greater than 2×10^{-3} Pa to yield a suitable replica. This shadowing is followed immediately by the deposition of a carbon support film as described previously for replica techniques. To achieve replicas with the highest resolution, the shadowing system should consist of tungsten/tantalum electron-beam guns, the specimen temperature should be held lower, and the vacuum should be increased.

The last step in the preparation process is cleaning the replica. The specimen with the replica is removed from the vacuum chamber, and the replica, usually with adhering specimen material, is floated off onto the surface of a water bath. The replica is transferred, via a loop or the like, to a series of digestive cleaning baths as described for replica techniques. After the final water wash, the replica is mounted on either a bare or a coated grid.

Interpreting Freeze-Etch Images Takes Practice Of primary consideration in interpreting a freeze-fracture replica is its overall quality. A properly shadowed platinum replica should have about a 2-nm to 3-nm point-to-point resolution. Somewhat higher resolutions can be attained in replicas made under ultrahigh-vacuum and ultralow-temperature conditions. In addition to high resolution, a good replica should have good contrast. Enough metal should have been deposited that even subtle topographical differences are visible; yet the metal coating should not be so thick that it obscures detail.

Other factors can affect replica quality. If the cleaning steps were insufficient to remove all traces of the specimen, the image will be blurred by these deposits. If the platinum replica shifts with respect to the carbon backing film, the image will have a peculiar ghost. Water condensing on the replica after platinum evaporation but before carbon evaporation can cause this phenomenon.

The quality of freezing and the environment between freezing and shadowing can affect image quality in several ways. First, if the freezing rate is too low, the membrane-bounded organelles take on a lumpy appearance, as opposed to the smooth contours observed when the growth of ice crystals is suppressed. The distribution of intramembranous particles should be uniform or in organized arrays, not patchy. If the ice around the specimen takes on a warty look, there may have been excessive water vapor in the system, which can also recondense on the specimen surface, adding a particulate artifacts.

Care should be exercised in interpreting the replica, even if it is generally free from physical artifacts. In freeze-fracture, the fracture plane tends to follow membranes and split them. If the specimen is etched, areas of the surface will be visible as well. Thus, there are four views of any membrane (Figure 6.17).

One of the unique and powerful characteristics of freeze-etch is that it enables the direct measurement of ultrastructural heights. Although freeze-fracture techniques are laborious and put a number of restrictions on the sample size, they offer a unique view of biological material and access to information that would be more difficult, if not impossible, to obtain by other techniques. Because the final TEM specimen image is formed in a completely different way, the freeze-fracture technique offers a good method for corroborating sectioning techniques as well.

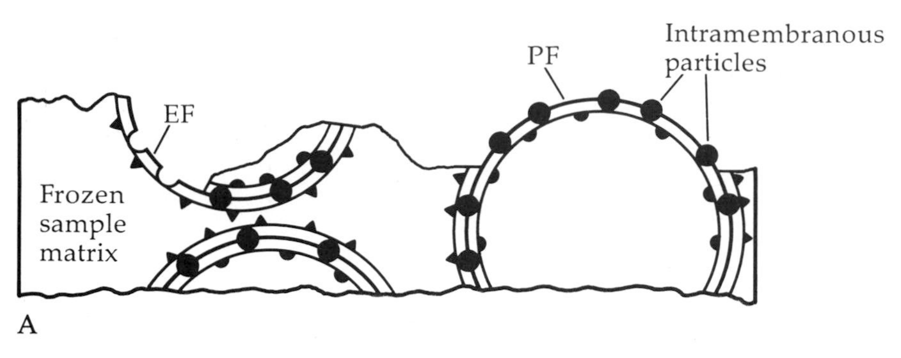

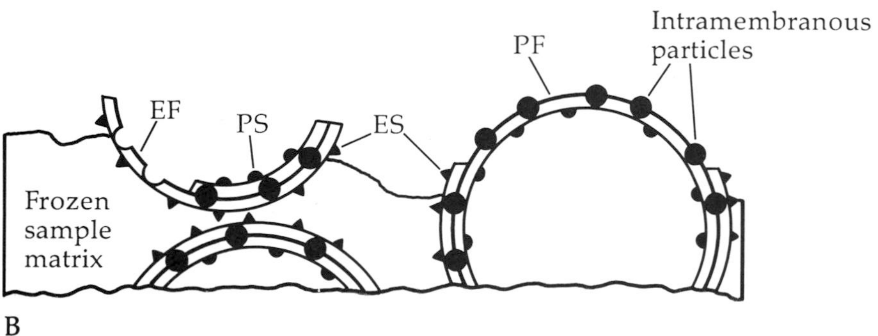

Figure 6.17 Membrane features revealed by freeze-fracture and freeze-etch. (A) Diagram of a freeze-fracture image. (B) Diagram of a freeze-etch image. EF: exoplasmic face (internal fracture plane). ES: exoplasmic surface. PF: protoplasmic face. PS: protoplasmic surface. Interpreting membrane faces, especially of organelles, requires practice. Freeze-etch is required for visualizing membrane surfaces.

CYTOLOGICAL TECHNIQUES

The TEM of well-preserved biological material has provided a corroborative and insightful addition to the biochemical study of living systems. Although still useful in explaining and comparing ultrastructural details, the TEM is increasingly important in subcellular biochemical analysis. Several techniques have been developed to characterize the chemical nature of the sample. Autoradiography, enzyme cytochemistry, immunological labeling, and X-ray microanalysis are widely used techniques. Each has a high degree of specificity and potential for quantitation.

X-ray microanalysis, as the name implies, involves the detection and analysis of X rays produced during beam-electron/sample interactions. This technique requires the use of a completely different signal transduction system than that of conventional (bright-field) TEM and is discussed in Chapter 8. Each of these techniques requires a number of preparation steps in addition to those required for general ultrastructural work. For successful TEM using these methods, it is essential that as much as possible is known about the integrity of the ingredients and the technical considerations involved. Testing the facility of the reactions involved at the light-microscopy level before undertaking the considerably more laborious TEM preparations is especially useful.

Autoradiographic Techniques

Autoradiography is the technique of using ultrathin sections of a radiolabeled specimen to produce a microradiogram in register with the section, allowing visualization of the locus of the introduced radiochemical in the specimen. EM autoradiography is a direct outgrowth of techniques developed for light microscopy, and differs from those techniques mainly in terms of temporal and spatial scale.

The first step in an autoradiographic study is to obtain as complete an understanding of the biochemistry of the system under observation as possible. Since the radioisotope must be introduced into living cells, the fate of the isotope-carrying molecule and possible pathways diverging from the incorporation target need to be known. As with any uniquely dangerous laboratory technique, proper training (and certification) and a means of disposing of the waste generated are requisite.

The radiographic process is essentially the same as the system of generating images in electron micrography. In the case of autoradiography, the electrons may have much higher energies than microscope-beam electrons have. Some of the emitted radiation interacts with silver halide crystals, applied in an emulsion in contact with the emitting site in the section, and sensitizes them. Subsequent development yields a stable silver grain in the proximity of the decay site of the radionuclide.

The choice of radionuclide is based on a number of factors. First, the isotope used should emit radiation with a low energy for maximum collection efficiency. High-energy particles are liable to pass through the thin emulsions required for good spatial resolution without sensitizing any silver atoms. Second, since a finite amount of radiolabeled chemical can be introduced into the system, a relatively short half-life is desirable because it allows for a high specific activity (i.e., millicuries/mMol). The half-life should not be so short, however, that the experiment (i.e., from synthesis of the radiolabeled compound to development of the radiogram) cannot be done. Generally a half-life of more than a month is required. Finally, the nuclide must be able to be incorporated into a biologically useful compound at a high enough concentration that a suitable exposure can be made from the minuscule amount in an ultrathin section. The primary choices are ^{14}C and ^{3}H. Because of the relative ease of tritiating a molecule as opposed to synthesizing it from scratch with ^{14}C, tritiated compounds are the most common.

After a suitably and stably labeled translocatable precursor has been acquired, the incorporation step is next. The specimen may be exposed to the labeled compound by injection, tissue incubation, or immersion. If the dynamics of the precursor movement within the cell are desired, the experiment can include a large chaser of unlabeled precursor to provide a sharp cutoff time for the label incorporation.

The fixation/dehydration/embedding regimen for labeled tissues is basically the same as for conventional ultrathin sections. The critical concern is that none of the steps leach out or destroy the target class of molecules. If the precursor is a protein, it may be advantageous to use an aldehyde (e.g., formaldehyde) that causes less crosslinking in its fixation than glutaraldehyde does to allow the unincorporated precursor to be leached out.

As is recommended for morphological studies, examining the tissue by light-microscopy techniques prior to making TEM preparations can save considerable time. Light-level autoradiograms should be checked to determine the prudence of pursuing the experiment at the EM level. Thick (1 μm to 2 μm) sections can be mounted on a gelatin-coated microscope slide, covered with radiographic emulsion, and exposed for a much shorter time than is required for the EM preparation, usually less than a month.

EM autoradiograms are prepared by either stretching a thin radiographic emulsion over thin (about 100 nm generally works well) sections mounted on a grid. Another approach is to mount thin sections on a slide coated with plastic film (Formvar or Parlodion), stain them for general morphology, coat them with a thin layer of carbon, and apply a layer of radiographic emulsion. This process is followed by allowing the incorporated radioisotopes in the section to expose the emulsion over a period of weeks in a cool dry environment. The exposure time can be approximated from the time for the light-level study. One approach is to increase the exposure time by a factor of ten for the EM preparations from tritiated materials.

After the requisite exposure time, the latent radiographic image can be developed in conventional film developers (e.g., D-19 or Kodak's Microdol-X). Usually the development time is short and after a brief acid stopbath, the emulsion is fixed in a nonhardening fixative such as pure sodium thiosulfate.

If the radiograms were made on a slide, the film is removed by gently immersing it in distilled water at an angle, and the whole film complex is floated off. Treatment with hydrofluoric acid (HF) may be necessary to remove some films, in which case several washes must be incorporated. Grids are then placed over the sections, and the film is picked up by blotting on paper.

Interpreting autoradiograms requires attention to numerous factors involving the chemistry of the system and the physics of the decay and radiographic processes. The resolution, for example, is dependent on how far the ionizing radiation travels laterally before striking a silver halide crystal. Even after sensitizing the crystal, the area on the crystal hit may be displaced from the eventual site of silver grain growth. The silver grains on the resultant micrograph thus represent points in circles of probability, rather than exact loci.

Immunological Techniques

Immunological techniques may be employed in TEM sample preparation of both sectioned and replica material or combinations of the two. Most often these techniques use specific immunoglobulins (e.g., IgG) or enzymes either directly coupled with colloidal gold particles or (in the case of the immunological techniques) coupled with colloidal gold that is coupled with secondary antibodies or protein-A, a bacterially derived polypeptide that reacts with many mammalian IgGs. These techniques, if applied with the appropriate controls, can allow the cytological localization and quantitation of very specific compounds. In addition, they can be sensitive enough that even a single target molecule can be visualized. The techniques involved are highly task-specific and may require modification of any or all of the preparation steps.

Immunoreagent Preparation Regardless of the type of TEM sample preparation, immunotechniques start with the acquisition of antibodies to the target compound. Acquiring antibodies involves isolating the chemical (or functional group) that is desired for labeling in the TEM preparation and making a mammal (or cell line) allergic to that compound. The result is the production of a group of serum proteins (antibodies) that have a strong binding affinity for specific chemical loci (epitopes) on the target molecule. In nature, this process helps the animal ward off invading pathogens; in vitro, it allows the researcher to harvest reagents with highly specific affinities. Effort spent here to assure the specificity and quality of the antibody is essential and cannot be overemphasized, since the immunolabeling process is time-consuming. Initial evaluation of the antibodies is much more efficient using enzyme-linked immunosorbent assay (ELISA), immunodiffusion, immunoelectrophoresis, radioimmuno assay (RIA), or other biochemical techniques, than using tissue thin sections.

Assuming that a sufficient quantity of the purified antigen can be acquired, polyclonal or monoclonal antibodies can be produced. Whole sera or immunoglobulin fractions may be used, or the antibodies can be purified further by affinity chromatography. Polyclonal and monoclonal antibodies to an increasing number of compounds are available from biochemical supply houses. Whatever the source and target, it is wise to be absolutely sure of the immunoreaction prior to embarking on TEM immunolabeling. In addition, experimental design should take into account that protein A does not bind to all types or species of mammalian antibodies; notable among these exceptions are IgM (initial response), goat, and rat antibodies. Protein

A also has weak or no affinity for some classes of IgGs (e.g., most IgG_1), an important factor if monoclonal antibodies are to be employed.

TEM Visualization of Immunoreaction Products Once the chemistry of the immunoreaction has been well characterized, the problem as to how to visualize this in situ in the TEM tissue preparation still exists. Since the biomolecules involved in the antigen/antibody reaction, large as they are, are still not resolvable in most (especially sectioned) TEM preparations, the usual approach to visualizing them is to incorporate an electron-dense marker in the reaction. At present the most commonly used marker is colloidal gold. These small globules of metallic gold can be purchased or made in the laboratory in diameters from a few nanometers to more than 100 nm. Preparations of fairly uniform size are easy to make, and differing antigenic sites can be labelled with different gold-labeled antibodies (multilabeling). Regardless of the size, colloidal gold appears as unambiguous black particles in the conventional TEM micrograph.

Colloidal gold can be bound directly to the antibodies for a specific antigen or to secondary antibodies to the primary antibody (e.g., goat antibodies to mouse antibodies). Another, easy approach is to use the binding of protein A with nonantigen-specific sites on the antibodies of most mammals. Protein A can be tagged with the gold and this combination bound to the antibody. Properly prepared and stored, the protein A/gold complex is quite stable in the buffer environments required for the antigen/antibody reaction.

Specimen Preparation for Gold-Labeling Since chemical fixation of specimens, by definition, involves reactions that alter the shape and reactivity of the biomolecules of the cells, care must be exercised in choosing fixation schedules. Quite often the standard dual fixation (aldehyde/osmium) eliminates the desired antigen/antibody reaction. Usually it is the secondary (osmium) fixation that destroys the reaction. This problem can be either avoided (i.e., by eliminating the secondary fixation) or reversed (by chemically removing the osmium from the sample surface). With the former approach, there is usually a trade-off in ultrastructural integrity, whereas the latter approach may not restore sufficient antigenicity to the section surface.

The choice of embedding medium is another factor that can influence the quality of immunolabeling on TEM thin sections. Although the embedding media of choice for morphological aspects, epoxies usually are the worst at preserving antigenicity of the specimen. Epoxy resins should not be completely discounted, however, because the effects of the embedment on antigenicity vary considerably. Another possibility is to use a modified acrylic, such as LR White or LR Gold. Modified acrylics generally are less damaging to antigenicity, and some may even be used at low temperatures, avoiding the possibility of thermally denaturing the antigenic sites. From the standpoint of ultrastructural preservation and sectionability, however, modified acrylics are often less desirable than epoxies.

Once the sample has been fixed and embedded, sectioning proceeds in much the same way as for morphological preparations. Better results are achieved if the section surface is smooth (with no knife marks, chatter, or dirt deposits) and the sections are picked up on fairly inert (nickel or gold) grids.

General Labeling Technique Applying the gold label is not difficult. The gold can be applied either to the surface of the specimen prior to fixation or to the sections after they are cut. The basic steps are similar to those for any specific antibody reaction. First, the nonspecific binding sites on the sample are blocked, usually with something like a buffered ovalbumin wash. Next, the antibodies are applied to the surface and are allowed to react with it for an empirically derived time (usually about an hour). Following the antigen/antibody reaction, the sections or sample surface are washed well with buffer and then reacted with the protein A/gold or secondary antibody-gold solution. After this final incubation, the grids are washed thoroughly to remove any unreacted reagents. Conventional staining with uranyl acetate and lead citrate can then be performed (Figure 6.18).

Controls are essential for immunogold labeling. There are three fundamental controls: 1) incubation with the probe alone (protein A/gold or secondary antibody/gold) to assess nonspecific binding of the probe to sample or to resin (this is particularly important if protein A is used, since it can bind to any endogenous sources of IgG in a specimen), 2) incubation with a primary antibody that has been exposed to an excess of antigen to confirm specificity, and 3) incubation with pre-immune serum or serum depleted of the specific antibody then probe to assess specificity of the IgG.

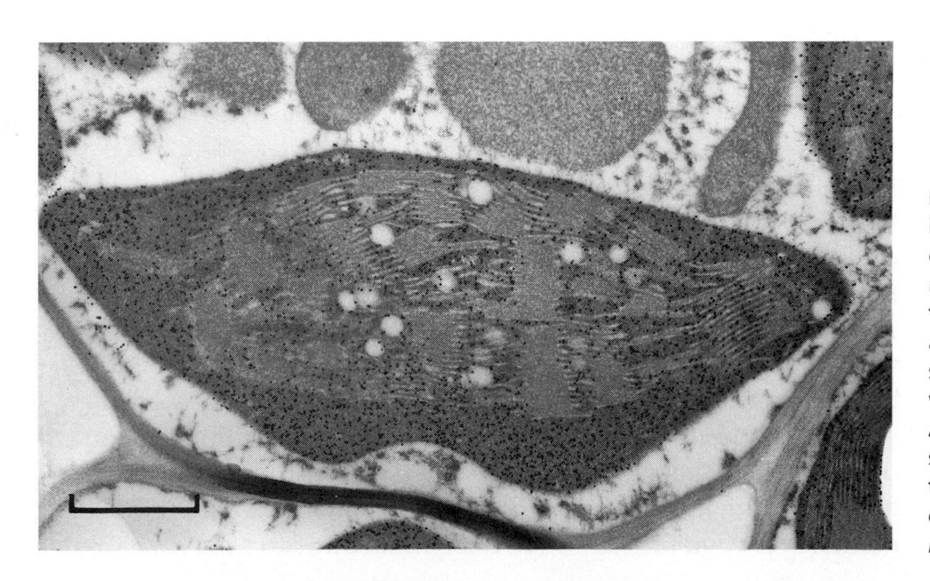

Figure 6.18 Protein A/gold immunolabeling. This ultrathin section through a celery leaf cell shows the localization of ribulose 1,5-bisphosphate carboxylase in the stroma of the chloroplast. The rabbit antibody was complexed to the antigenic sites on the section, followed by reaction with colloidal gold bound to protein A. After immunolabeling, the section was stained to highlight the morphology of the labeled structures. Bar = 1 μm. Photo courtesy of J. D. Everard and B. A. McFadden.

Labeling Freeze-Fracture Replicas Although most immunolabeling of TEM samples is done on sectioned material, useful data can be gained by the labeling freeze-fracture replicas. Whereas complete digestion of the biological material from a replica is considered a standard practice for freeze-fracture morphological studies, a number of immunolabeling techniques can be devised, based on the retention of sufficient biological material to preserve the antigenic sites. Surface antigens on cells or organelle preparations can be tagged with the immunogold label. In addition, face antigenic sites can be labeled by lightly shadowing the fractured surface of a freeze-fracture sample with platinum and then can be reacted with immunogold. Although still quite new, these techniques are opening up exciting possibilities for correlating structure and function on a molecular level using TEM.

Interpreting Labeled TEM Preparations Great strides have been made in the last decade in the ease and reliability of immunological labeling for electron microscopy. Resolving gold particles of any size is not difficult because of their high electron-scattering power. The most important operational consideration is often the specificity of binding of the gold, however ligated, to the target antigen. Concurrent preparations with the appropriate control conditions, mentioned earlier, are crucial to an unequivocal study.

If the experiment is designed properly, it can permit the localization, qualification, and quantitation of specific biochemical entities at a resolution unattainable by other means.

PREPARATION OF NONBIOLOGICAL MATERIALS

Nonbiological and materials-science samples may require techniques other than the three broad approaches discussed earlier in this chapter. For some applications, however, the approaches used in biology can be modified. Conversely, some materials-science preparation techniques may be useful in preparing especially refractory biological materials, such as bone.

Ultrathin sectioning, most commonly used with biological samples, may also be applied to a variety of materials-science samples. Ultrathin sectioning of polymer samples generally is easy. If the mechanical properties between the components differ greatly, cryoultramicrotomy can often help tremendously by stiffening the less rigid components to near their glass-transition temperature. Ultrathin sections can be used to examine the sizes, distribution, and boundaries between phases or structures in the material.

If the electron density of the sample is relatively uniform, the different components and structures often can be contrasted in a manner analogous to the staining used for biological samples. Composite materials bearing double bonds can be stained with osmium tetroxide

(OsO_4) vapor simply by incubating the sections for about an hour, on a dry support and in a sealed vessel containing several drops of OsO_4 solution. Ruthenium tetroxide (RuO_4), which reacts with both aromatic and nonaromatic unsaturated domains, can be useful as a stain for composites with an aromatic component, such as styrene.

Ultrathin sectioning is not limited to organic samples and is being applied in a widening sphere of materials research endeavors. Ultrathin sectioning is a useful method for preparing thin-film cross sections. Layered metal films can be embedded in a resin support, just as biological samples are, then ultrathin sectioned. Often the main differences between preparing a biological sample and a materials-science sample are the details of the technique rather than the approach. For some samples, for example, a metal with a low melting point, such as indium or gallium, may be a better support than a resin system used in biology; the sectioning equipment and procedures are similar to those for biological samples.

Chemical polishing or electropolishing techniques have been the most common methods for thinning inorganic materials for TEM. With chemical polishing techniques, the samples are immersed in a chemical solution that gradually removes material from the surfaces uniformly and smoothly enough for TEM. Usually an oxidative chemical (such as perchloric acid or nitric acid) is compounded with other modifying agents into a solution that allows the removal of surface material to occur without etching or pitting the surface. For conductive samples the results of the thinning effect can be greatly enhanced by applying an electrical potential to the system, with the specimen as an anode, and running the process at lower temperatures. The final and most common elaboration is to thin the sample by directing a stream of electrolytic solution at the sample surface through a cathodic nozzle (jet polishing).

Ion thinning (also called ion milling) is another way of rendering a specimen thin enough for TEM imaging. For some brittle nonconducting materials (such as minerals and ceramics), ion thinning is the only method that produces a usable thin sample with a minimum of deformations. It can also be used to clean the surfaces of samples prepared by other methods. Ion thinning is particularly useful for two-phase specimens that thin at different rates when polished electrolytically (although some multiphase materials may ion-thin differentially).

Ion thinning involves directing a 4-keV to 6-keV beam of argon (or other inert gas) ions at both sides of a specimen. As the ions strike the specimen, the surface atoms are sputtered out, thinning the specimen. A specimen stage rotates the specimen during thinning; its angle of incidence to the beams may be set up between 5° and 30°, the optimum angles for sputtering. The rate of sputtering varies with the composition and structure of the specimen, the energy of the ion beam, and the angle of the beam as it bombards the specimen. Ion-thinning devices (also called ion mills) generally have a diffusion- or turbomolecular-pumped vacuum system (see Chapter 3), which is used to remove room air and scavenge the gas introduced in the ion streams (see Goodhew (1973) and Sawyer and Grubb (1987) in Further Reading for details on this and other methods).

As with the other specimen preparation techniques discussed in this chapter, ion thinning requires experimentation and practice with each type of specimen to choose the correct variables. Inorganic samples are no less prone to preparation artifacts than biological specimens are, although the causes may be different. Artifacts from stresses, deformations, and thermal and chemical alterations can be introduced at most stages of sample preparation. Paramount is the fidelity of representation of the final specimen sample in the TEM, or at least recognition of the artifacts.

FURTHER READING

Aldrich, H. C., and W. J. Todd, eds. 1986. *Ultrastructure Techniques for Microorganisms.* Plenum, New York. (Preparation of bacteria, viruses, and nucleic acids; immunolabeling)

Bullock, G. R., and P. Petruz, eds. 1989. *Techniques in Immunocytochemistry.* 4 vols. Academic Press, New York. (Fixation and labeling techniques for immunolabeling)

Goodhew, P. J. 1973. *Specimen Preparation in Materials Science.* North-Holland, Amsterdam. (Ion thinning)

Hayat, M. A. 1975. *Positive Staining for Electron Microscopy.* Van Nostrand Reinhold, New York. (Positive staining for ultrathin sections)

Hayat, M. A. 1989. *Principles and Techniques of Electron Microscopy,* vol. 3. CRC Press, Boca Raton, FL. (General specimen preparation)

Hayat, M. A., ed. 1991. *Colloidal Gold: Principles, Methods, and Applications.* 3 vols. Academic Press, San Diego. (Immunolabeling for TEM and SEM)

Hayat, M. A., and S. E. Miller. 1989. *Negative Staining.* McGraw-Hill, New York. (Negative staining)

Polak, J. M., and I. M. Varnell, eds. 1984. *Immunolabelling for Electron Microscopy.* Elsevier, Amsterdam. (Immunolabeling)

Rash, J. E., and C. S. Hudson. 1979. *Freeze Fracture: Methods, Artifacts and Interpretations.* Raven, New York. (Introduction to freeze-fracture)

Robards, A. W., and U. B. Sleyter. 1985. *Low Temperature Methods in Biological Electron Microscopy.* Elsevier, New York. (Cryogenic preparation techniques)

Rogers, A. W. 1973. *Techniques of Autoradiography.* Elsevier, Amsterdam. (TEM and light autoradiography)

Sawyer, L. C., and D. T. Grubb. 1987. *Polymer Microscopy.* Chapman and Hall, London. (Ion thinning)

Sommerville, J., and U. Scheer, eds. 1987. *Electron Microscopy in Molecular Biology.* IRL Press, Washington, DC. (Preparation of nucleic acids, chromatin, autoradiography, immunolabeling)

7

SPECIMEN PREPARATION FOR SEM

Preparing specimens for SEM is often less complex and time-consuming than it is for TEM because whole samples, instead of ultrathin sections, are examined. The ability to examine many different sample types in the SEM has stimulated the development of a vast array of sample preparation methods. This chapter covers many of these methods; the Further Reading section provides sources for additional details.

Nearly all samples require some type of preparation before examination in the SEM. Samples must be (1) devoid of water, solvents, or other materials that could vaporize in the vacuum and either contaminate the column or cause vacuum problems, (2) firmly mounted, and (3) electrically conductive. Most biological samples do not meet any of the three criteria, and therefore, require extensive preparation. Most metals and semiconductors meet criteria 1 and 3 and need only to be mounted. Nonconductive, nonbiological samples such as plastics, composites, and ceramics meet criterion 1 only and need to be mounted and made electrically conductive.

MOUNTING

Most SEM samples are mounted on metal holders called stubs. Stubs come in a variety of designs and sizes and are usually made of aluminum (Figure 7.1). Most SEMs accept only a specific type and size of stub.

The samples must be fastened to the stub by a mounting medium; glues and tapes are most common. The mounting material should have the following characteristics: (1) stability under bombardment by the beam of electrons, (2) noninterfering appearance or morphology when present in an SEM micrograph, (3) minimal release of air or solvents (outgassing) in the vacuum of the column, (4) mechanical stability, (5) acceptable chemical element background if used for x-ray analysis, and (6) good electrical conductivity, if possible, to help reduce problems of charging.

Many glues, over one hundred types, have been described as useful in mounting SEM samples. In addition to the characteristics for mounting materials mentioned previously, drying time and viscosity are important criteria. If the glue is extremely thin and watery, some specimens will act as a sponge and absorb so much glue that it may obscure features of interest. In general then, the glue should be fairly viscous. Too short a drying time may not allow sufficient time to manipulate and mount the sample properly. If the drying takes too long, the sample may be more likely to absorb the glue. In addition, the glue may not be completely dry when the sample is finally placed in the SEM.

The fast-curing epoxy resins make excellent glues. They are viscous, dry quickly, have good mechanical strength, are beam stable, and have minimal outgassing. However, they are nonconductors. Silver glue and graphite cements are somewhat less viscous than epoxy resins, but they share many of the desirable properties of epoxy resins and have the added advantage of being conductive.

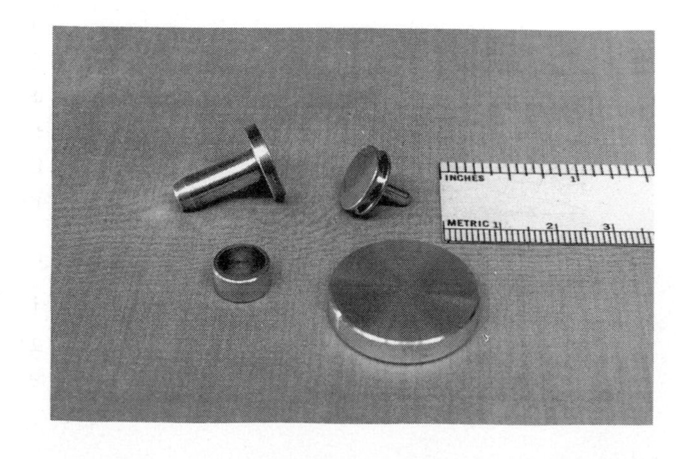

Figure 7.1 Stubs used to mount SEM samples come in a variety of sizes and shapes. Different SEMs and experiments require different styles.

Various types of tape have been used for mounting SEM samples, and they have some distinct advantages over glues. Because they are not liquid, the sample is never damaged by absorption. In addition, the sample may be moved around and manipulated with more ease. Tapes are especially useful in mounting small particles, such as chemical crystals or pollen, because the samples can be dusted onto the tape; usually a high percentage of the particles adhere. Some tapes are electrically conductive with a conductive adhesive (e.g., a copper or aluminum backing with a conductive adhesive). By fastening the backing side to the stub with a conductive glue, the conductive adhesive on the tape may be left face up for sample placement. Special adhesive tabs are also available that are used in a manner similar to that of double-sided tape but that outgas less.

Mounting on a stub may be undesirable for certain types of samples, such as forensic samples, samples in which other analysis techniques are to be used after SEM examination, and rare or valuable samples that might be damaged or altered by the glue or adhesive. Some samples, such as pieces of metal or coins, may be placed directly on the sample holder without using a stub. Such samples usually have enough mass that movement is not a problem. Small vises that replace the sample holder are available and may be used to hold samples that require a firm mount.

A

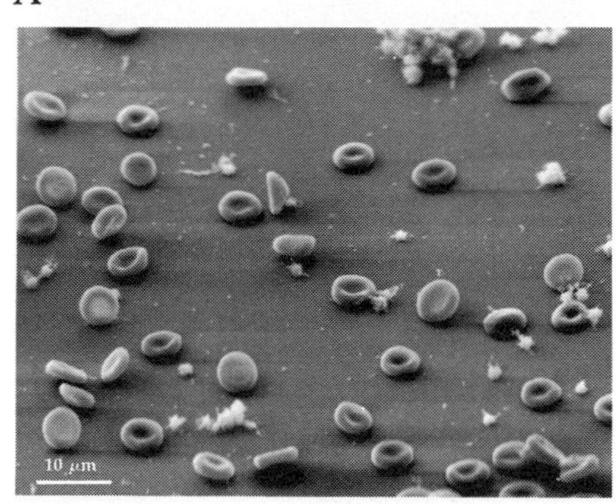

B

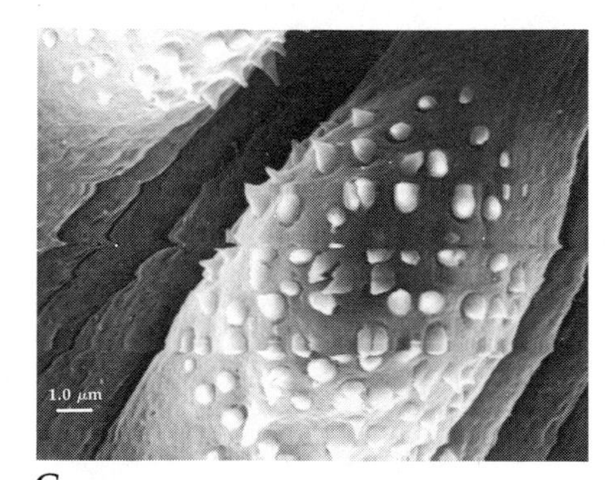

C

Figure 7.2 Charging of samples in the SEM. (A) A sample of fly ash, the airborne byproduct of a coal-fired electrical generating facility. The type of charging is abnormal contrast. The larger particles are too bright, the smaller particles too dark. (B) Human red blood cells illustrating the type of charging that produces lines. (C) The surface of a schistosome, a type of trematode worm that causes a severe disease endemic to parts of Asia, Africa, and South America. Beam-deflection charging is shown.

COATING FOR CONDUCTIVITY

A sample placed in the SEM usually must be conductive. Metals are conductive, and most semiconductors present no problems. Samples that are nonconductive or that have nonconductive portions, however, may cause severe problems during the imaging and photographic processes.

When a nonconducting portion of a sample is viewed in the SEM, a negative charge builds up gradually on that area from bombardment by the beam of electrons. A buildup of negative charges on the sample may produce one or more of three conditions commonly referred to as charging (Figure 7.2): (1) lines on the screen or photograph, (2) abnormal contrast, or (3) breaks in the image that appear as if the image has been split. The lines are caused by the negative charge on the sample deflecting the secondary electrons emitted from the sample. Their deflection causes a sudden change in the secondary-electron signal level at a given location. This change appears as a dark line on the image if there is a decrease in signal or a light line on the image if there is an increase in signal. Abnormal contrast results from uneven distribution of the negative charge on the sample because the more negatively charged areas produce more secondary electrons than the areas with less negative charge do. The breaks, or splitting of the image, are caused by a very high negative charge on the sample, which deflects the beam of electrons, resulting in a sudden shift so that the image is generated from different areas.

Although charging can produce problems in all imaging modes, it is especially a problem in the normal secondary-electron mode, since the low energy of secondary electrons causes them to be deflected easily. The backscattered-electron mode is far less susceptible to charging because of the high energy of backscattered electrons.

Most Samples Are Coated with Gold in a Sputter Coater

Most nonconductive SEM samples are coated with a very thin layer of metal to make the surface conductive and to prevent the formation of negative charge. The most common method of coating a sample is to apply gold using a sputter coater. The sputter-coating process became prevalent in the early 1970s as its advantages became apparent and as commercial equipment became available.

A sputter coater is a small apparatus that can fit on a counter (Figure 7.3). It requires a two-stage mechanical pump, which is normally mounted on the outside of the coater. A small vacuum chamber contains a ring-shaped magnet covered with a sheet of gold, or in some models the gold is a concentric ring between the poles. The gold

A

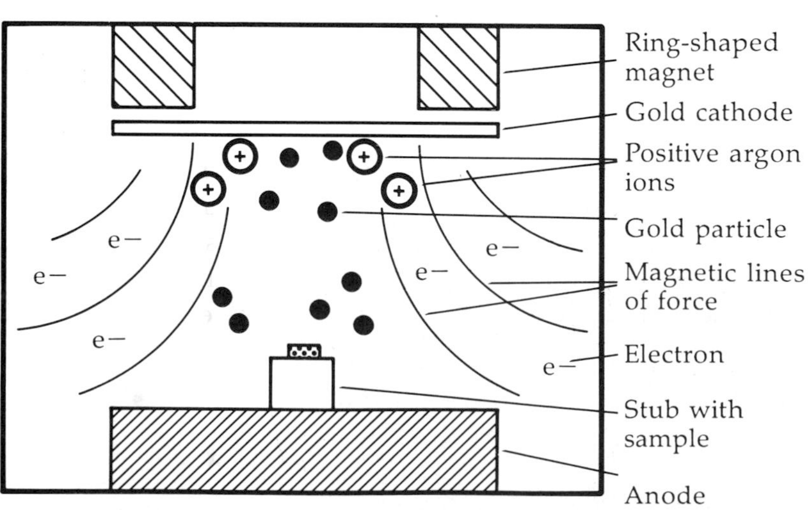

B

Figure 7.3 Sputter coater. Most SEM samples are coated with gold in a sputter coater. (A) A typical sputter coater. (B) Schematic of a sputter coater. Argon ions strike the gold cathode, ejecting gold atoms that coat the sample.

functions as the cathode, and a metal plate at the bottom of the chamber acts as the anode. The mounted samples are placed on the anode.

During operation, the chamber is pumped to a low-level vacuum. Argon gas is then allowed into the chamber at a controlled rate resulting in a decrease in vacuum and a constant purge of the gas. A voltage differential of 100 V to 2,000 V applied between the anode and the cathode ionizes the argon into positive argon ions and free electrons. The electrons are accelerated toward the positive anode (where the samples are placed), and the argon ions are accelerated toward the negative cathode. When the argon ions strike the gold cathode, the kinetic energy of the collision is imparted to the atoms in the gold lattice structure, which results in the ejection of gold atoms and electrons. The gold atoms undergo multiple collisions with the argon ions, after they are ejected from the cathode and, as a result, strike the sample from all directions.

The electrons from the cathode normally would be directed toward the anode, where many of them would strike the sample, causing heating, which could be detrimental to the specimen. The magnet produces magnetic lines of force, however, that deflect the electrons given off by the cathode so that they are directed toward the outside of the chamber and away from the samples. In addition, the electrons move in a spiral manner because of the magnetic field, which results in a retardation of the electrons and an increased path length. The increased path length gives an increased argon ion yield and, therefore, an increase in sputtering efficiency.

The coating obtained from a sputter coater is very uniform with few shadow effects because it comes from a broad metal plate and because the gold atoms collide with the argon atoms. The thickness required to make the sample conductive varies somewhat with the texture of the sample, but 10 nm to 30 nm is usually adequate.

The coating of gold obtained from a sputter coater consists of discrete particles of gold of a distinct size. Most modern sputter coaters produce a particle size less than 3 nm in diameter, a dimension that is less than the resolution of most SEMs. Increased emphasis is being place on smaller particle size of sputtered gold as the resolution limit of SEMs is improved. In the future, there may be other methods of metal coating, such as ion beam sputtering and electron beam evaporation, that can achieve smaller particle size.

Samples for X-Ray Analysis or Backscattered-Electron Imaging Are Coated with Carbon

Samples that are to be examined using backscattered electrons or X-ray analysis should not be coated with a heavy metal, such as gold, because it will obscure the atomic number differences for the backscattered electrons and will absorb the X-rays from the sample. Carbon coating is used instead because carbon has a very low absorption factor for X rays and does not obscure the atomic number differences needed for backscattered electrons.

One of the best methods of carbon coating is to use a carbon string evaporator. The device is similar to a sputter coater in appearance and operation. The samples are placed in a small vacuum chamber. A piece of carbon fiber string is stretched between two support posts at the top of the chamber and the chamber is then pumped to a low vacuum using only a rotary mechanical pump; no argon is used. At operating vacuum, a current is suddenly applied to the string, causing it to flash, or evaporate, very quickly. The carbon atoms collide with each other and with the remaining air molecules in the chamber and, as a result, coat the sample from all directions. The thickness of the coating may be varied by varying the diameter of the string.

Some Nonconductive Samples May Be Examined without a Conductive Coating

Nearly all nonconducting samples are coated with metal or carbon for conductivity, but there are two situations in which a coating may not be desirable. When a protocol for sample analysis requires initial examination in the SEM followed by analysis of the same sample using other methods, subsequent analysis using almost any analytical method would be hindered by the coating. A procedure does exist for coating the sample with silver, which can then be removed using photographic chemicals. The photographic chemicals, however, may cause deleterious effects on some samples. A second situation in which a coating is not desirable is with any sample for forensic analysis because jurisprudence usually dictates that forensic samples not be modified in any manner.

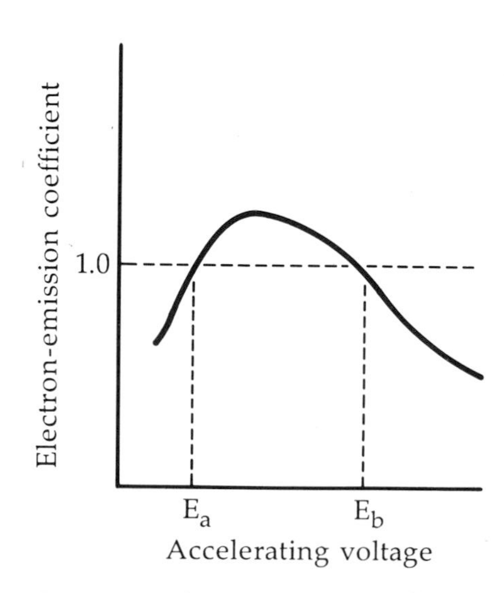

Figure 7.4 Electron-emission coefficient as a function of accelerating voltage. Samples may be examined at accelerating voltages between E_a and E_b with no charging artifacts.

Most nonconducting samples can be examined uncoated with no charging artifacts, if the proper accelerating voltage is selected, a process that requires knowledge of the electron emission coefficient. The electron-emission coefficient is the sum of all secondary electrons, Auger electrons, and backscattered electrons emitted from the sample divided by the sum of the electrons in the electron beam. A plot of electron-emission coefficient as a function of accelerating voltage for a nonconducting sample is shown in Figure 7.4. At very low accelerating voltages (less than E_a), the coefficient is less than one; fewer electrons are being emitted from the sample than are entering, and therefore there is a buildup of negative charge on the sample. At intermediate accelerating voltages (between E_a and E_b), the coefficient is greater than one; more electrons are being emitted from the sample, and therefore there is a buildup of positive charge on the sample. At

A

B

C

Figure 7.5 Uncoated sample of table salt examined at various accelerating voltages. (A) When examined at 1 kV, charging artifacts are present. (B) At 5 kV, no charging artifacts are present. (C) At 15 kV, charging artifacts are present.

higher accelerating voltages (above E_b), there is again a buildup of negative charge. If an accelerating voltage between E_a and E_b is selected properly, an uncoated sample may be examined with no negative charging artifacts (Figure 7.5). A positive charge is present on the sample at accelerating voltages between E_a and E_b. The effect of a positive charge on the sample, however, is only to bring about a slight reduction in secondary electron emission, resulting in minimal effect on the image. The values for E_a are in the range of several hundred volts and the values for E_b are in the range of 1 kV to 5 kV. The actual values vary with the type of sample and may also vary from one point on the sample to another.

SPECIAL METHODS FOR VARIOUS SAMPLE TYPES

Replicas Are Often Useful

In some situations placing the sample in the SEM is not possible—e.g., if the sample is too large to fit inside the SEM but cannot be cut into smaller pieces because it is too valuable or it is unique. Another example is with the study of the growth of a plant stem, in which the same cells are to be observed at intervals. In general, any study in which micrographs at intervals are needed or when the sample cannot be destroyed is a case in which replica methods can be used.

Replica methods have been used for some time for examining materials-science samples or geological samples, which often are too large for SEM examination. Replica techniques using dental-impression materials have produced excellent results with a variety of biological and nonbiological samples; reproduction of surface features at a magnification of several thousand times can be obtained. A negative replica of the sample, made with the dental-impression material, is filled with epoxy resin. The resin is cured and forms the positive replica, which is then mounted, coated, and examined in the SEM.

Metallurgical Samples May Require Polishing

The SEM can be very useful for examining metallurgical specimens. Dendritic structure, grains, twins, and the constituents of alloys can be observed (see Figure 5.14). Such samples usually require polishing and sometimes etching. Plastics, silicon crystals, silicon chips, coal, ceramics, and multilayered materials may also require polishing.

The methods for polishing are standard metallurgical procedures. Initially, the sample may be cut and rough-polished on abrasive paper. Final polishing consists of using abrasive powders with progressively finer grit on a polishing wheel. Some samples may also require a technique called electropolishing. Finally, some metallurgical samples may require the use of etching chemicals to show grain boundary structure. (See the Further Reading section for additional information on these procedures.)

Cryogenic Techniques Can Be Used to Examine Many Special Sample Types

Many samples cannot be fixed chemically and examined in the SEM. Such samples include oils, grease, certain adhesives, wet chemical mixtures, and various foodstuffs like ice cream and whole milk. Standard sample preparation methods do not work with these samples, and any of these samples would contaminate the SEM column and cause severe vacuum problems. In other situations, chemical fixation methods work but have deleterious effects on the sample. For example, plant leaves may be prepared using standard methods, but the wax structure of the leaves is usually changed or destroyed. If an environmental SEM is not available, cryogenic SEM techniques offer a very useful alternative for special samples. By maintaining the sample in the SEM chamber at a sufficiently low temperature (approximately –130°C or lower), no vaporization of water or other organic materials occurs.

Two components are required for examining frozen of samples in the frozen state: a preparation unit that allows freezing and coating of samples, and a device that will keep the sample frozen when viewed in the SEM. Several manufacturers now offer cryogenic units.

BIOLOGICAL SAMPLE PREPARATION

Most biological samples can be chemically fixed in a solution, dehydrated, critical-point dried, mounted, and coated with gold.

Fixation

The principles of fixation for SEM the same as those for TEM (see Chapter 6). Specimens for SEM may require cleaning because, unlike with TEM, surfaces are usually the area of interest. It is generally desirable to have one dimension be 2 mm or less so that fixation and dehydration solutions can penetrate the tissue. The other dimension can be quite large, up to 10 mm or 15 mm. Primary fixation is usually sufficient for many samples. Secondary fixation in osmium tetroxide (OsO_4) may slightly reduce charging problems (see the section on gold coating earlier in this chapter) and may be needed for some isolated cells to avoid membrane holes.

Dehydration

Dehydrating the sample in a solvent is necessary because the water used in the fixative and buffer wash is not miscible with the liquid CO_2 used in the critical-point drying procedures (described in the next section). Many solvents are miscible with both water and liquid CO_2.

Ethanol and acetone are the most common solvents. Both work well although ethanol has some advantages for SEM bulk samples: (1) It is less toxic, (2) it is less volatile so is less likely to allow the sample to dry during specimen transfer into the critical-point dryer, (3) it does not extract bound lipids as much as acetone does, and (4) it causes less swelling at the beginning of dehydration and less shrinkage at the end. Usually rather large steps of 25%, 50%, 75%, 95%, and 100% ethanol or acetone in water are used with about 10 to 20 minutes in each step at room temperature. Swelling that occurs in most tissue as dehydration is initiated is offset somewhat by shrinkage as the higher concentrations of solvent are approached. Several changes in 100% solvent are recommended to ensure complete dehydration.

Critical-Point Drying

If a sample were air-dried in the natural state, after immersion fixation or after solvent dehydration, the surface tension forces of the water or solvent as it evaporated from the cells would cause extensive distortion. A critical-point dryer is a device for drying a sample in the absence of surface tension forces. It consists of a metal chamber with thick walls designed to withstand high pressures and has an inlet valve, an outlet valve, a removable cover, and usually a viewing window (Figure 7.6). A CO_2 siphon-type cylinder is attached to the inlet. The siphon- type cylinder has a pipe that draws liquid CO_2 from the bottom, rather than gaseous CO_2 from the top.

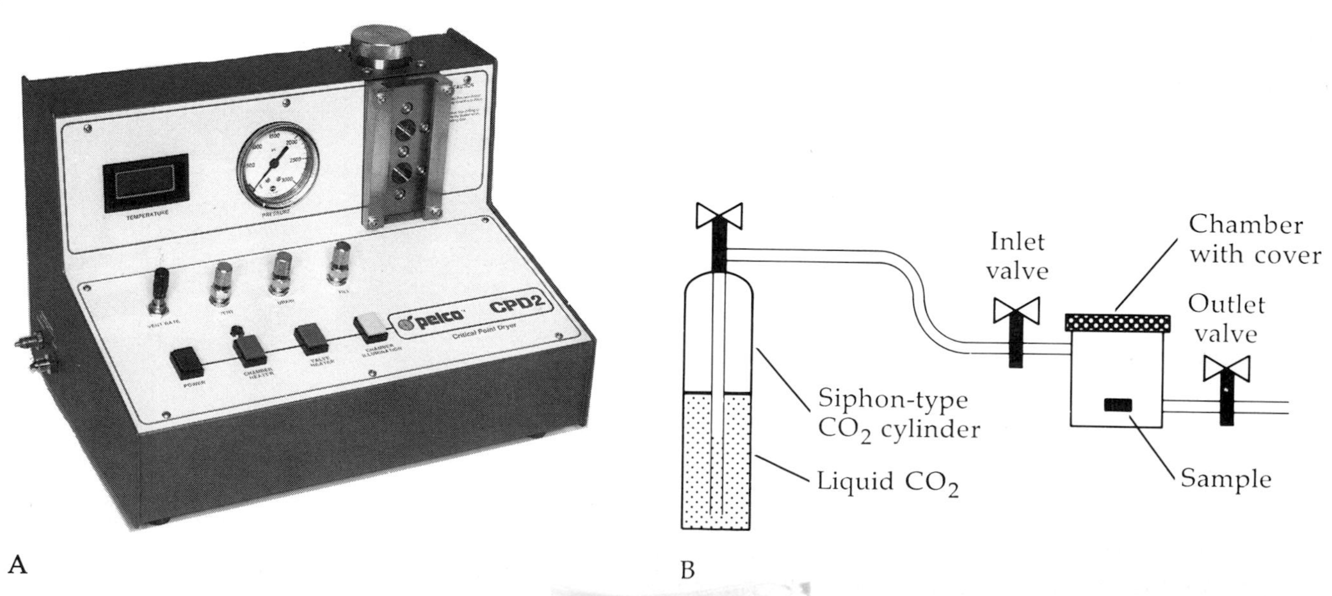

Figure 7.6 Critical-point dryer. (A) Critical-point dryer. Several different designs are available from different manufacturers. (B) Schematic of a critical-point dryer. During critical-point drying, the dehydrated sample is infiltrated with liquid CO_2 in the chamber. As the temperature increases, the pressure increases, finally driving all of the liquid into the gas phase, at which point the sample is dry and the gas can be exhausted from the chamber.

The critical-point drying process begins by cooling the chamber to around 5° to 10°C to ensure that the liquid CO_2 remains a liquid when it is admitted to the chamber. The samples are inserted into the chamber and the chamber cover attached. The chamber is then filled with liquid CO_2. The liquid CO_2 gradually replaces the solvent in the sample, and a mixture of solvent and liquid CO_2 forms in the chamber. To reach the goal of having only liquid CO_2 in the chamber and in the specimen, the chamber is usually flushed several times by opening the outlet and inlet valves for brief periods.

The next step is the heating phase. The liquid level in the chamber is adjusted to approximately one-half full, and both the inlet and the outlet valves are closed. (If the chamber were completely full of liquid, the pressure would rise too rapidly in relation to the temperature, a phenomenon referred to as "going around the critical point" that is thought to cause distortion of the sample.) After the liquid level is adjusted, the chamber is heated, which causes the liquid phase to become less dense as the molecules move farther apart and the gas phase to become more dense because it is a closed system. At the critical point, the density of the liquid phase becomes equal to the density of the gas phase. The sample is now held within a very dense gas. It is dry and has not been distorted because of the absence of surface-tension forces at the critical point. The critical point for CO_2 is 31.3°C at 75.5 kg/cm^2 pressure (1,072 psi). Usually the temperature and pressure are allowed to rise somewhat above these values to ensure that condensation does not occur.

The exhaust phase begins after the desired temperature and pressure have been attained. The outlet valve is opened slowly, and the pressure is allowed to drop gradually to zero over a period of several minutes to prevent the sample from exploding. The critical-point drying process is complete at this point. To prevent absorption of moisture, the samples should be stored in a vacuum chamber with a desiccating agent.

Liquid CO_2 is used as a transitional fluid because its temperature and pressure at the critical point are within a range that does not damage biological material. In theory, a sample could be critical point dried in water, however, the critical point of water, 374°C at 223.9 kg/cm^2 (3,184 psi), would destroy the tissue. Liquid CO_2 has acceptable values at the critical point.

Critical-point drying of tissue fixed by immersion is the most common method of preparing biological samples for the SEM. It is not without problems and artifacts however. One major artifact is that of tissue shrinkage. A linear shrinkage of 25% is not uncommon. With some tissues, such as lung, it may be as high as 50%. The values tend to be less, sometimes much less, with samples that have a hard exoskeleton or cell wall, such as insects and plants. Secondary fixation with osmium tetroxide secondary fixation has been shown to reduce the shrinkage somewhat.

ALTERNATIVE METHODS FOR BIOLOGICAL SAMPLES

A number of alternative methods of sample preparation in certain circumstances may be preferable over those already mentioned.

The type of sample and experimental conditions may require that chemical fixation followed by critical-point drying, replica formation, or cryogenic SEM be avoided.

Air-Drying

A very old and easy technique is to air-dry the sample and then mount it and coat it with metal. Air-drying was used in the early days of SEM with poor results. Most samples became extremely distorted and shrank. The method is excellent in certain instances in which the sample has a large amount of internal strength or hardness. Some examples are teeth, bones, pollen, exoskeletons, wood, shells, kidney stones, and plant stems with high silicon content, etc.

Drying from a Solvent

A variation of the air-drying procedure is to air-dry the sample from a solvent and then mount and coat it. Most commonly used solvents, such as alcohol or acetone, have surface-tension forces much lower than those of water, so there may be less distortion than from air-drying alone. Overall preservation, however, is usually far inferior to preservation with critical-point drying.

Vapor Fixation

Vapor fixation has special value in preparation of samples that can be damaged by solutions, such as certain fungi that produce long chains of spores, in which the chains are usually broken when the samples are processed using solutions of fixatives and dehydrating agents. Another example involves determining the types of pollen adherent to various insect parts. The solutions in a normal fixation could wash away many of the pollen grains.

Vapor fixation consists of placing the sample in a small, closed container with a small amount of aqueous osmium tetroxide (1% to 4%). A glass petri dish with a small watch glass of solution makes a simple, effective vapor chamber. The samples are placed in the covered dish and are allowed to stand, covered for 72 to 96 hours in a fume hood. Osmium tetroxide (OsO_4) produces very powerful, penetrating vapors that interact with the surface of the sample to fix, harden, and stabilize it. Following fixation, the sample is removed, air-dried, mounted, and coated. The success of vapor fixation depends on the type of tissue being fixed. It is often necessary to accept some loss of quality in the fixation to be able to observe the structures of interest. For instance, in the example of the fungus, the hyphal cells may be partially collapsed, while the chains of spores are intact.

Vapor Prefixation

A variation of vapor fixation is to use a brief vapor prefixation (approximately 30 minutes) in OsO_4 vapors followed by a normal fixation (by immersion in glutaraldehyde), dehydration, critical-point drying, mounting, and coating. This technique has resulted in greater retention of fungal structures on leaves and has been shown to be of value in other cases as well.

Freeze-Drying

Freeze-drying a sample produces excellent results in some situations. The sample is maintained in a frozen state under a high vacuum, during which sublimation (the direct change from solid to vapor without going through a liquid state) occurs. Surface-tension forces thus do not act on a sample as they would if it were air dried. A sample can be freeze dried either from water or from solvents.

Freeze-drying has advantages over the normal fixation and critical-point drying process: (1) Shrinkage of animal tissue may less, (2) waxy or cuticular coatings of plants are preserved much better, but only if freeze-drying is not done from solvents, and (3) the chemical elements within a cell are not altered as much, which makes freeze-drying a useful method of preparing samples for X-ray analysis (Chapter 8).

Freeze-drying also has some disadvantages: (1) The formation of ice crystals may result in both internal and external damage to cells, (2) preservation of morphological features is usually no different from that obtained with critical-point drying, (3) the process is more difficult because the techniques and equipment are not as standardized as they are for critical-point drying, and (4) the process requires substantially more time. For these reasons, freeze-drying is not as popular as is critical- point drying in the preparation of SEM samples.

Poly-L-Lysine Procedure

Small particulates, especially those in solution, can be difficult to prepare for SEM. Examples include blood, bacteria, protozoans, fungal zoospores, and organelle suspensions. One method of preparing such samples is to use poly-L-lysine as a biological glue to coat glass coverslips. The particulates are allowed to settle on the coated coverslips, which are then processed, mounted, and coated. The adhesive effect is believed to be due to the negative charge on the cell surface combined with the positive charge of the poly-L-lysine. Particulates not in suspension can also be processed with this method. For example, pollen may be fixed in suspension and then treated as a particulate. The advantages of this technique are that cells are retained well and are usually well dispersed, and the coverslips produce a nice smooth background for photographs.

Hydrated Analysis

Some samples can be examined successfully in their normal hydrated state in a conventional SEM. Plant materials tend to give better results than most samples because the cell wall provides rigidity and protection to minimize shriveling and to prevent water loss in the vacuum. This method violates some of the rules of electron microscope operation because wet samples normally are not placed in an EM. Because of the high-vacuum environment, wet samples increase the potential for contamination of the column. Most samples cannot be examined in the hydrated state. Use of an environmental SEM (see Chapter 5), rather than a conventional SEM, however, enables the examination of nearly any wet sample.

Wet samples should be mounted quickly, glues should be conductive, and only small amounts of glue should be applied. Following mounting, the samples usually need to be sputter- coated. The water present in the tissue imparts some conductivity. Unless very low accelerating voltages are used in the SEM, though, most samples will be charged if not coated. The samples should then be examined in the SEM and photos taken as quickly as possible. The sample quickly loses water and begins to collapse or shrivel in the high-vacuum conditions. The examination time varies with the type of specimen and its particular state. Total times of between two and ten minutes are common.

A major disadvantage of this method is that it causes contamination of the SEM column and of the SEM vacuum system. Certain types of plants with very high wax content, such as cabbage, can cause severe contamination problems. Labs that do hydrated analysis usually have to clean the SEM column every few days. The manufacturer's service representative should be consulted before this type of examination is attempted, because in many situations it voids warranties, thus excluding repairs and cleaning from service contracts.

Chemical Conductivity

An alternative to coating samples with metal is to make the sample conductive chemically; no metal coating is then needed. The tissue of the sample is made conductive by impregnating tissue with osmium tetroxide (OsO_4), which is reacted chemically with either thiocarbohydrazide or tannic acid to form a conductive complex that is retained within the tissue. The nature of the chemical reaction has been described as a mordant or a ligand, and many variations of it exist.

One advantage of this noncoating method is that it may provide images that, on some sample types, show fine structure better than those obtained with metal coating. A second advantage is that fewer charging problems may occur with some sample types because the tissue of the sample itself is conductive.

Cryostat Preparation

Often it is desirable to look at the inside of an organ, organism, or structure (e.g., the internal structure of a kidney, a mushroom, or the bud of a flowering plant) using the SEM. The sample must be dissected to expose the area of interest. Ordinary hand dissection with a razor blade or a scalpel often results in damage to the sample because of uneven cutting, tearing, or shredding of the surface. An alternative is to use a cryostat, a microtome contained in a refrigerated chamber.

Samples can be mounted whole on a holder in the cryostat and then frozen. Once the sample is frozen, it can be sectioned. Unless they are necessary to determine the depth of sectioning, the sections may be discarded. The cryostat, then, is used to make a precise, clean cut at a desired location in a sample. After sectioning is complete, the whole nonsectioned portion of the sample is removed, placed in a buffer to thaw, rinsed several times to remove remaining embedding material, and then fixed, dehydrated in a solvent, critical-point dried, mounted, and coated. This method is especially useful when it is necessary to section large samples while maintaining some control over the plane of sectioning.

Cryofracturing

Cryofracturing is a very useful technique in preparing small samples in which the fine details of internal structure are to be studied. Fixed and dehydrated samples are frozen in liquid nitrogen and are cut with a cooled razor blade. The blade does not actually cut the sample, but rather fractures it, thereby producing a very clean surface, free of many artifacts associated with simply cutting samples at room temperature with a razor blade. The method usually works best when the sample is fractured after dehydration but before critical-point drying. Cryofracturing is an easy technique; its only major drawback is that precise control over the location of the fracture is difficult. With homogeneous samples, this is not a problem. With other samples it may be necessary to use alternative methods, such as resin embedding and removal or cryostat preparation.

Resin Embedding and Removal

Occasionally it is necessary to view the internal structures of a very small sample. Cutting the sample with a razor blade could produce artifacts and would not be precise enough needed. Use of a cryostat may not give the precision needed on very small samples, and cryofracturing gives very little precision of the plane of cut or fracture. If the sample is fixed and embedded in an epoxy resin, as is done with TEM samples, it can be mounted in an ultramicrotome, and thick 1 μm to 4 μm sections can be made using the precision mechanical and viewing devices inherent in an ultramicrotome. The unsectioned portion is placed in special solvents to extract the resin. The tissue is then rinsed in ethanol or acetone, critical-point dried, mounted, and coated for SEM examination. An additional use or benefit of this method is that thick sections can be obtained, stained, and viewed in a light microscope, or ultrathin sections can be obtained, stained, and viewed in a TEM. This flexibility allows correlation of SEM, TEM, and light microscopy for a single piece of tissue. As an alternative, the thick sections can also be viewed in the SEM, as described in the next section.

Viewing Sections in the SEM

The SEM may be used to view thick sections of embedded material with resolution between that of the light microscope and that of the TEM. A single block of tissue may be used to prepare sections for the SEM, TEM, and the light microscope, and thus, to correlate the information obtained from the three instruments. In addition, serial sections can be examined in the SEM, and the same sections can then be viewed in a light microscope. The thick sections in resin are stained with both uranyl acetate and lead citrate, mounted on a grid or carbon stub, coated with carbon, and examined in the SEM. In general, sections of 200 nm or greater thickness should be used in order to give sufficient contrast in the SEM.

Intracellular Structures May Be Observed

The SEM usually is thought of as an instrument for observing the external features of cells or tissues and the TEM as an instrument for observing the internal components of cells. The SEM, however, can provide significant information on internal components, if proper preparation procedures are used. One such technique is called the Tanaka procedure. A key feature of this method is that it removes cytoplasmic material from the cell by prolonged exposure to dilute osmium tetroxide (OsO_4) and thus allows the internal organelles to be seen. The complete process consists of fixation in (OsO_4) or aldehydes, followed by infiltration with dimethyl sulfoxide (for its cryoprotective action), followed by freezing in liquid nitrogen and cracking of the tissue on a cold metal block. The sample is then soaked in dilute OsO_4 for 24 to 72 hours to remove cytoplasm, and then treated with tannic acid followed by OsO_4 to increase conductivity. Finally, the sample is dehydrated, critical-point dried, coated with metal in a high-resolution coating apparatus, and observed in the SEM.

SEM HISTOCHEMISTRY FOR BIOLOGICAL SAMPLES

Histochemistry (or cytochemistry), the chemistry of the cell, normally refers to techniques that enable identification of the composition or the function of cellular components.

The ability to relate structure to function can be extremely useful. SEM histochemistry can be a powerful tool because it combines the advantages of the SEM (e.g., depth-of-field, high and low magnification) with the localization properties of the histochemical procedures.

Imaging

Histochemical procedures have been used in light microscopy for many years, and hundreds of procedures exist in the literature, most of which rely on the deposition of a colored dye as the end product. The colored dye is then observed with the light microscope. TEM histochemical procedures rely on the deposition of a metal as the end product, which, being electron-opaque, is therefore, dark on the TEM screen. Visualization of heavy-metal end products in an SEM usually requires the use of backscattered electrons because of their ability to show differences in atomic number. Because they originate from deeper within the sample, using backscattered electrons enables observation of stained intracellular detail (Figure 5.13). A significant disadvantage of using backscattered electrons is that the resolution is much lower than it is with secondary electrons.

X-ray analysis can also be used to detect the heavy-metal histochemical end products. X-ray analysis has the advantage enabling positive identification of the element in the end product. The backscattered-electron image cannot identify the end product; rather it must be assumed that because areas appear brighter, they are a result of the metal end product. Thus, it is wise in any histochemical procedure to use both backscattered-electron and X-ray analysis. A disadvantage of X-ray analysis is that a true image cannot be produced based on X rays; dot maps (see Chapter 8) may simulate an image, however. In addition, the resolution of X rays is lower than that of backscattered electrons.

Staining

Staining methods deposit a metal on the sample for a specific function (e.g., to localize a type of organelle or an enzyme or to identify the chemical nature of a deposit). Examples are procedures that deposit a silver end product to detect the presence of polysaccharides, procedures that deposit silver to stain nuclei and nucleoli, and procedures that deposit a number of heavy metals to localize peroxidase or phosphatase activities.

Autoradiography

Autoradiography is a technique for localizing substances labeled with radioactive isotopes. Cells are allowed to take up the radioactive label and then are coated with a liquid photographic emulsion. Silver grains, which are activated by the radioactive label, remain as metallic silver after processing. The specific deposition of the silver can then be correlated with the position of the original labeled substance. SEM autoradiography has not been used extensively, although there is considerable potential for development. (See Chapter 6 for details of its use in TEM.)

Immunocytochemistry

Immunocytochemistry is very useful for the SEM. The technique is used to localize specific cell-surface target molecules (antigens, lectins, enzyme substrates) with specific identifier molecules (antibodies, polysaccharides, enzymes) that recognize and attach to the target. The identifier molecules have attached visualizers (gold, latex spheres), which are then observed in the SEM. The technique used with SEM has been used extensively with animal cells and yeasts and, to a limited extent, with plant cells. Colloidal gold as a label is probably the most common visualizer, and extensive literature exists concerning its use. The gold is observed using the backscattered-electron image or, in some instances, using the secondary-electron image. Latex spheres can be used as a visualizer and have the advantage of being very recognizable in using the secondary-electron mode. Latex spheres and labeled colloidal gold are available from a number of EM suppliers. (See Chapter 6 for details of the use of gold in TEM.)

FURTHER READING

Barnes, S. H., and S. Blackmore. 1986. Plant ultrastructure in the scanning electron microscope. *Scanning Electron Microsc.* 1986(1):281–289. (Intracellular structures)

Becker, R. P., and J. S. Geoffroy. 1981. Backscattered electron imaging for the life sciences: Introduction and index to applications—1961 to 1980. *Scanning Electron Microsc.* 1981(4):195–206. (Histochemical stains)

Beckett, A., and N. D. Read. 1986. Low-temperature scanning electron microscopy. In *Ultrastructure Techniques for Microorganisms.* H. C. Aldrich and W. J. Todd, eds. Plenum, New York. (Cryogenic techniques)

Boyd, A. 1974. Histological and cytochemical methods for the SEM in biology and medicine. In *Scanning Electron Microscopy,* by O. C. Wells. McGraw-Hill, New York. (Cryofracture)

Boyd, A. 1978. Pros and cons of critical point drying and freeze drying for SEM. *Scanning Electron Microsc.* 1978(2):303–314. (Freeze-drying)

Boyd, A., E. Bailey, S. J. Jones, and A. Tamarin. 1977. Dimensional changes during specimen preparation for SEM. *Scanning Electron Microsc.* 1977(1):507–518. (Critical-point drying)

Boyd, A., and E. Maconnachie. 1981. Morphological correlations with dimensional change during SEM specimen preparation. *Scanning Electron Microsc.* 1981(4):27–34. (Drying from a solvent)

Echlin, P. 1981. Recent advances in specimen coating techniques. *Scanning Electron Microsc.* 1981(1):79–90. (Sputter-coating)

Echlin, P. B., B. Chapman, L. Stoter, W. Gee, and A. Burgess. 1982. Low voltage sputter coating. *Scanning Electron Microsc.* 1982(1):29–38. (Sputter-coating)

Flegler, S. L., and K. K. Baker. 1983. Use of scanning electron microscopy in plant pathology. *Scanning Electron Microsc.* 1983(4):1707–1718. (Histochemical stains, autoradiography)

Flegler, S. L., and G. R. Hooper. 1980. Ultrastructure and development of *Nidularia pulvinata. Mycologia* 72:472–482. (Cryostat)

Gabriel, B. L. 1985. *SEM: A User's Manual for Materials Science.* American Society for Metals, Metals Park, OH. (Polishing and etching methods)

Gamliel, H. 1985. Optimum fixation conditions may allow air drying of soft biological specimens with minimum cell shrinkage and maximum preservation of surface features. *Scanning Electron Microsc.* 1985(4):1649–1664. (Air-drying, drying from a solvent)

Goldstein, J. I., D. E. Newbury, P. Echlin, D. C. Joy, C. Fiori, and E. Lifshin. 1981. *Scanning Electron Microscopy and X-ray Microanalysis.* Plenum, New York. (Uncoated samples)

Hizume, M., S. Sato, and A. Tanaka. 1980. A highly reproducible method of nucleolus organizing regions staining in plants. *Stain Technology* 55(2):87–90. (Histochemical stains)

Hodges, G. M., A. W. Carbonell, M. D. Muir, and P. R. Grant. 1974. Uses and limitations of scanning electron microscope autoradiography. *Scanning Electron Microsc.* 1974(1):159–166. (Autoradiography)

Hodges, G. M., J. Southgate, and E. C. Toulson. 1987. Colloidal gold—a powerful tool in scanning electron microscopy immunocytochemistry: An overview of bioapplications. *Scanning Microsc.* 1:301–318. (Immunocytochemistry)

Horiguchi, T., F. Sasaki, H. Takahama, and K. Watanabe. 1984. Identification of cells by backscattered electron imaging of silver stained bulk tissues in scanning electron microscopy. *Stain Technol.* 59:143–148. (Histochemical stains)

Humphreys, W. J., B. O. Spurlock, and J. S. Johnson. 1974. Critical point drying of ethanol-infiltrated, cryofractured biological specimens for scanning electron microscopy. *Scanning Electron Microsc.* 1974:275–282. (Cryofracture)

Klomparens, K. L., S. L. Flegler, and G. R. Hooper. 1986. *Procedures for Transmission and Scanning Electron Microscopy for Biological and Medical Science.* Second edition. Ladd Research Industries, Burlington, VT. (Poly-L-lysine, resin embedding and removal)

Klomparens, K. L., M. A. Peterson, D. C. Ramsdell, and W. G. Chaney. 1986. The use of scanning electron microscope autoradiography to localize the blueberry shoestring virus in its aphid vector. *J. Electron Microsc. Technique* 4:47–54. (SEM autoradiography)

Krause, C. R., J. M. Ichida, and L. S. Dochinger . 1986. Osmium vapor pretreatment of Gnomia infected leaves. *Scanning Electron Microsc.* 1986(4):975–978. (Vapor prefixation)

Ledbetter, M. C. 1976. Practical problems in observation of unfixed uncoated plant surfaces by SEM. *Scanning Electron Microsc.* 1976(2):453–460. (Hydrated analysis)

McCormack, S. M., F. J. Tormo, and J. D. B. Featherstone. 1991. A straightforward scanning electron microscopy technique for examining non-metal coated dental hard tissues. *Scanning Microsc.* 5:269–272. (Uncoated samples)

Mills, A. A. 1988. Silver as a removable conductive coating for scanning electron microscopy. *Scanning Microsc.* 2:1265–1271. (Removable silver coatings)

Molday, R. S. 1976. Immunolatex spheres as cell surface markers for scanning electron microscopy. *In Principles and Techniques of Scanning Electron Microscopy. Biological Applications*, vol. 5. M. A. Hayat, ed. Van Nostrand Reinhold, New York. (Immunocytochemistry)

Murakami, T. N. I., T. Taguchi, O. Ohtani, A. Kikuta, A. Ohtsuka, and T. Itoshima. 1983. Conductive staining of biological specimens for scanning electron microscopy with special reference to ligand-mediated osmium impregnation. *Scanning Electron Microsc.* 1983(1):235–246. (Noncoating techniques)

Murphy, J. A. 1980. Non-coating techniques to render biological specimens conductive/1980 update. *Scanning Electron Microsc.* 1980(1):209–220. (Noncoating techniques)

Murphy, J. A. 1982. Considerations, materials, and procedures for specimen mounting prior to scanning electron microscopic examination. *Scanning Electron Microsc.* 1982(2):657–696. (Mounting)

Nockholds, C. E., K. Moran, E. Dobson, and A. Phillips. 1982. Design and operation of a high efficiency sputter coater. *Scanning Electron Microsc.* 1982(3):907–915. (Sputter-coating)

Pameijer, C. H. 1979. Replication techniques with new dental impression materials in combination with different negative impression materials. *Scanning Electron Microsc.* 1979(2):571–574. (Replicas)

Quattlebaum, E. C., and G. R. Corner. A technique for preparing *Beauveria* spp. for scanning electron microscopy. *Canad. J. Bot.* 58:1700–1703. (Vapor fixation)

Sargent, J. A. 1988. Application of cold stage scanning electron microscopy to food research. *Food Microstruct.* 7:123–135. (Cryogenic SEM in food science)

Scala, C., G. Cenacchi, P. Preda, M. Vici, R. P. Apkarian, and G. Pasquinelli. 1991. Conventional and high resolution scanning electron microscopy of biological sectioned material. *Scanning Microsc.* 5:135–145. (Viewing sections)

Soligo, D., E. Pozzoli, M. T. Nava, N. Polli, G. Lambertenghi-Deliliers, and E. de Harven. 1983. Cytochemical methods for the backscattered electron imaging mode of scanning electron microscopy: Further applications to the study of human leukemic cells. *Scanning Electron Microsc.* 1984(4):1795–1802. (Histochemistry)

Spicer, S. S., B. A. Schulte, and J. D. Shelburne. 1983. Carbohydrate cytochemistry by transmission and scanning electron microscopy. *Scanning Electron Microsc.* 1983(4):1827–1834. (Histochemistry)

Tanaka, K., and A. Mitsushima. 1984. A preparation method for observing intracellular structures by scanning electron microscopy. *J. Microsc.* 133:213–222. (Intracellular structures)

Winborn, W. B. 1976. Removal of resins from specimens for scanning electron microscopy. In *Principles and Techniques of Scanning Electron Microscopy. Biological Applications*, vol. 5. M. A. Hayat, ed. Van Nostrand Reinhold, New York. (Resin embedding and removal)

Witcomb, M. J. 1981. The suitability of various adhesives as mounting media for scanning electron microscopy I: Epoxies, sprays and tapes. *J. Microsc.* 121:289–308. (Mounting)

Witcomb, M. J. 1984. The suitability of various adhesives as mounting media for scanning electron microscopy. II. General purpose glues. *J. Microsc.* 139:75–114. (Mounting)

8

X-Ray Analysis

X rays are produced in all electron microscopes (SEM, TEM, and STEM) because of specimen–beam interactions. Detection and analysis of the X rays with accessory attachments allow determination of the presence, amount, and distribution of the elements in the sample. The advantages of X-ray analysis are that it does not destroy the sample and that it is done on a spatial basis; the composition of one part of a sample may be compared with that of an adjacent part.

The ability of X-ray analysis to differentiate composition on a spatial basis is probably its greatest advantage over other methods of elemental analysis. The analytical spatial resolution (i.e. the ability to differentiate the separate locations of two or more elements) is not as high as the imaging resolution of an SEM, TEM, or STEM, however. For instance, the resolution of a TEM might be 0.2 nm, whereas its analytical spatial resolution might be only 0.1 μm. The resolution of an SEM might be 4 nm, and its analytical spatial resolution only 5 μm with a metal sample and 30 μm with a biological sample.

X rays are produced from the entire area of specimen–beam interaction, as are secondary electrons, backscattered electrons, and

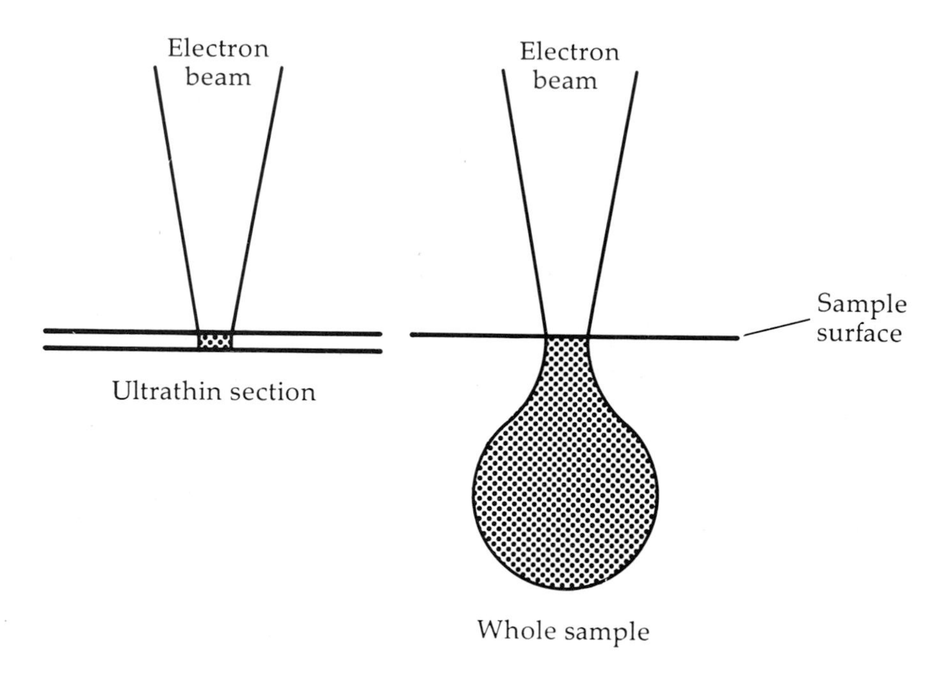

Figure 8.1 Specimen–beam interactions in thin sections and whole samples. The analytical spatial resolution is better in thin sections than in whole samples because of the decreased width of specimen–beam interactions.

Auger electrons. X rays are photons, however, rather than electrons, and have a much lower probability of losing their energy because of inelastic scattering. The escape depth of X rays is thus essentially the same as that of the entire area of specimen–beam interaction. The greater analytical spatial resolution in a thin section is due to the decreased width and depth of the area of specimen–beam interaction (Figure 8.1).

X-ray analysis is not as sensitive as many other analytical techniques, such as neutron activation, atomic absorption, and inductively-coupled plasma spectroscopy, which have detectability limits measured in parts per million or better. The detectability limit of X-ray analysis is measured in parts per thousand. A typical value of concentration of a given element that can be detected for a whole sample in an SEM is 0.08% by weight and for a thin section in the TEM is 0.004% by weight. The increased sensitivity with a thin section is due to less absorption of the X rays and decreased detection and production of background X rays.

X-RAY PRODUCTION AND NAMING

X rays are a result of inelastic scattering (see Chapters 4 and 5). A vacancy in an inner orbital shell is filled by an electron from a shell of higher energy. The energy difference between the shells may be emitted in the form of an X ray. The energy and the wavelength of the X ray produced are characteristic of the element and are related by the equation $\lambda = 1.2398/E$, where λ is the wavelength in nanometers, and E is the energy in keV. The actual energy of the X rays may vary 2 to 3 eV due to the Heisenberg Uncertainty Principle.

Each X ray produced has a name based on the name of the shell (K, L, M, N) in which the vacancy was created and on the number of orbital shell jumps made by the electron that filled the vacancy. Hence, a one-shell jump is denoted by a subscript α, a two-shell jump by ß, a three-shell jump by γ, and so on. For example, a vacancy in the K shell filled by an electron from the M shell would create a $K_{ß}$ X ray. Because of the presence of electron subshells, electron spin differences, and quantum energy levels (i.e., that only certain transitions are possible between subshells), naming can be very complex; for example, there are $L_{\alpha1}$, L_{III1}, $M_{III\gamma}$, etc. In practice, the X rays most commonly used in analysis are $K_{\alpha1}$, $K_{ß1}$, $L_{\alpha1}$, $L_{ß1}$, and $M\alpha$.

An element often produces more than one type of X ray because of multiple beam electrons striking the sample, each of which may produce a different type of interaction, and because of a cascading effect of interactions. For instance, a vacancy in the K shell might be filled by an electron from the L shell, producing a $K_{\alpha1}$ X ray; then the vacancy in the L shell might be filled by an electron from the M shell, producing an $L_{\alpha1}$ X ray, and so on.

Each type of X ray is called a line. If enough X rays of a given line are generated, they produce an X-ray peak in the spectrum of X rays analyzed. Each X-ray line has a characteristic energy and wavelength. The X rays from a given shell are referred to collectively as a family (i.e., the K family line, the L family line, etc.).

The beam of electrons must have a certain energy, called the critical excitation energy, to remove an electron from a given shell; if the energy is not sufficient, no X-ray lines from that shell (or subshell) will be produced. The energy of the beam is determined by the accelerating voltage of the electron microscope. For instance, silver has a K-shell critical excitation energy of 25.517 keV. If the accelerating voltage is 25 kV, no K-line X rays will be produced; if the accelerating voltage is 26 kV, $K_{\alpha1}$ and $K_{ß1}$ X rays will be produced with characteristic energies of 22.162 keV and 24.942 keV respectively. The critical excitation energy is always slightly higher than the characteristic energies. In the example just described, K-line X rays would be produced with an accelerating voltage of 26 kV; but the quantity produced would be small. To produce a quantity of X rays sufficient for an accurate analysis, excess voltage must be used; normally an accelerating voltage of one-and-a-half to three times the critical excitation energy is considered sufficient.

Critical excitation energy differences between elements lead to differences in the analytical spatial resolution when analyzing for these

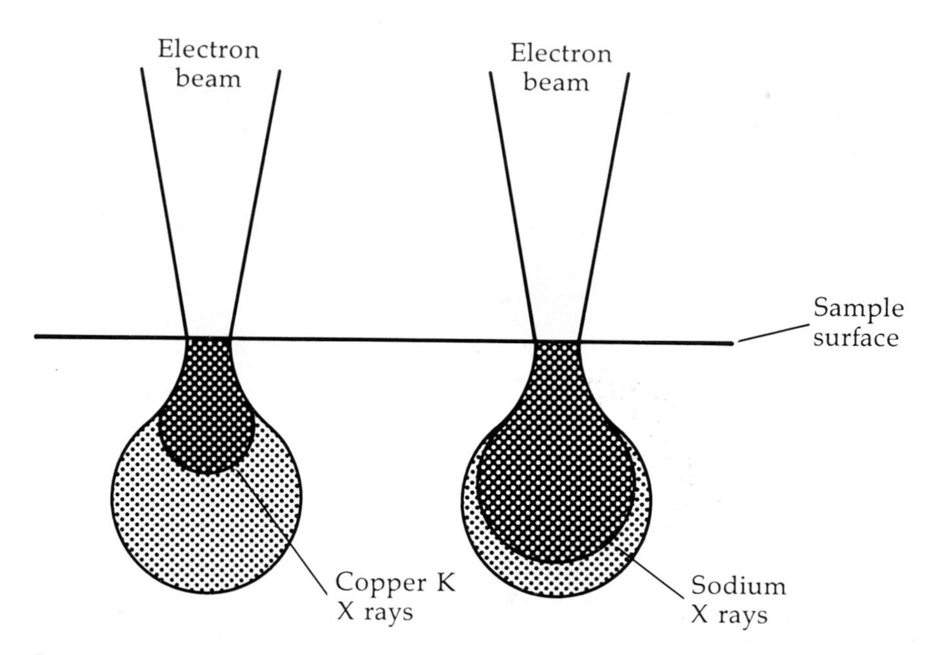

Figure 8.2 Analytical spatial resolution in whole samples. The analytical spatial resolution in a whole samples can vary with the element and the specific X-ray line being detected.

specific elements. For instance, the critical excitation energy for the sodium K shell is 1.08 keV, and that for the copper K shell is 8.98 keV. When incident electrons enter the sample, they gradually lose energy over distance because of scattering. The greater the distance, the more energy is lost. X rays can be produced by beam electrons from sodium that have traveled a much greater distance than those of copper because of the lower critical excitation energy of sodium compared to that of copper (Figure 8.2). Differences between the analytical spatial resolution of the various elements are more pronounced in whole samples than they are in thin sections because in whole samples the beam electrons eventually lose all their energy, whereas in thin sections only a small percentage of the energy is lost. Therefore, the greater variation in energy of beam electrons in whole samples produces major differences in analytical spatial resolution for different elements.

In addition to the production of characteristic X rays (i.e., X rays in which the energy or wavelength is characteristic of the element), noncharacteristic X rays are produced. These X rays are variously called background X rays, continuum X rays, or bremsstrahlung (braking radiation) X rays and are produced as a result of deceleration of the beam electron when it enters the coulombic field between the nucleus and the tightly bound electrons. The background X rays range in energy from zero to the energy of the electron beam. A normal spectrum has characteristic X-ray peaks superimposed on the background X rays.

MEASURING THE ENERGY AND WAVELENGTH OF X RAYS

The elements in a sample can be determined by measuring either the energy or the wavelength of the X rays produced. The measurement of energy is called energy-dispersive spectroscopy (EDS), and the measurement of wavelength is called wavelength-dispersive spectroscopy (WDS).

EDS has several advantages over WDS. The total cost of an EDS system may be one-fourth that of a WDS system. With EDS, it is possible to determine in one spectrum the presence of all the elements in the area being analyzed. With WDS, it is usually possible to detect no more than four elements at one time. Each detector must be adjusted to the correct wavelength for each element separately, and most systems have no more than four detectors. A complete search of the periodic table of elements might take several minutes with EDS and several hours with WDS. EDS has greater sensitivity at low beam currents because more of the X rays produced are detected. With the low beam currents that must be used for many polymer and biological samples, sensitivity can be improved by up to five times. WDS has a higher sensitivity with high beam currents. Many polymer and biological samples would be destroyed by the high beam current.

The most important advantage of WDS over EDS is that of spectral selectivity—i.e., the ability to separate the X rays of two elements whose energies, and therefore wavelengths, are close to each other. Spectral selectivity typically is 10 eV for WDS, compared to 135 eV for EDS. Peak overlaps often occur in EDS. For instance, a common overlap encountered in EDS is that between calcium and potassium. The potassium K_β peak (3.589 keV) overlaps with the calcium $K\alpha$ peak (3.690 keV) because the separation of the lines is only 101 eV; the analysis of calcium is impossible with high amounts of potassium and low amounts of calcium, unless other approaches are taken (see Peak Intensity Measurement and Peak Deconvolution in this chapter). If the analysis were being done with WDS, no overlap would exist in this example.

Because of its advantages, EDS is much more common than WDS. An EDS system is often attached to an SEM, TEM, or an STEM (Figure 8.3). A WDS system may be attached to an SEM but is more often purchased on an SEM specially designed for WDS, called a microprobe. WDS systems are seldom attached to TEMs or STEMs. Because EDS is much more popular, it is the only system described here in detail.

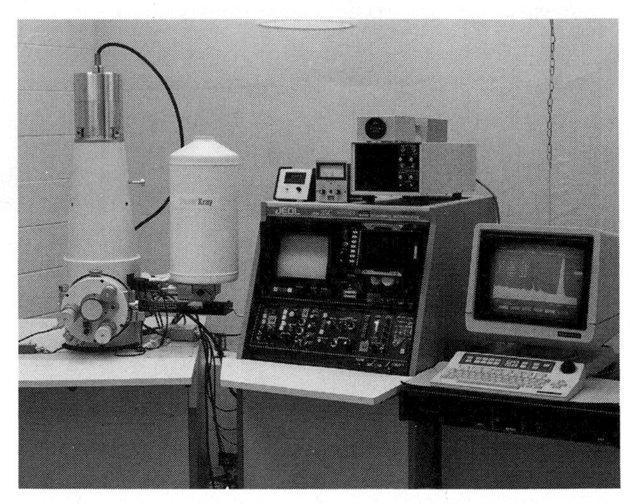

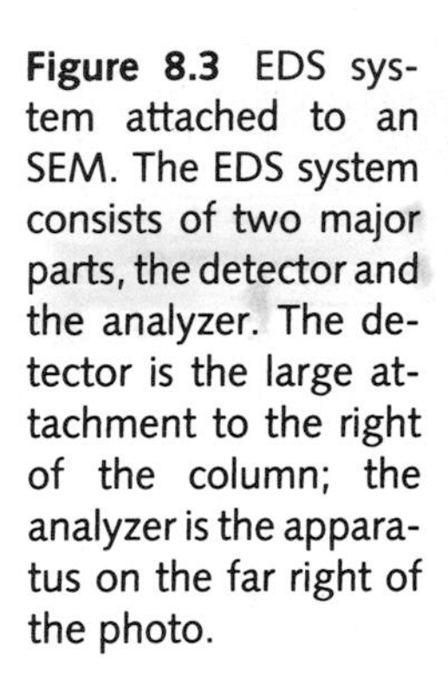

Figure 8.3 EDS system attached to an SEM. The EDS system consists of two major parts, the detector and the analyzer. The detector is the large attachment to the right of the column; the analyzer is the apparatus on the far right of the photo.

CONSTRUCTION OF THE EDS DETECTOR

The EDS detector attaches to the column of an electron microscope and consists of a number of components: collimator, window, detector crystal, field-effect transistor, and liquid nitrogen dewar (Figure 8.4).

The collimator is located at the front portion of a metal tube on the bottom of the detector. The collimator reduces stray radiation by physically blocking its entrance and by virtue of its walls, which are lined with carbon; X rays and backscattered electrons not entering in straight paths strike the carbon and are thus absorbed.

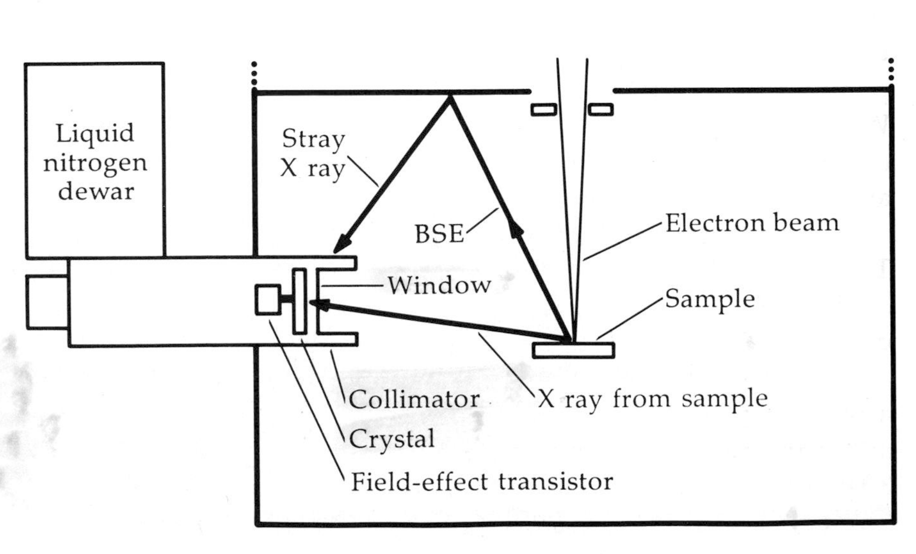

Figure 8.4 Schematic of an EDS detector attached to an SEM. The main components of the detector and the mechanism of X-ray detection are illustrated. (The components are not drawn to scale with reference to each other.) The collimator reduces stray radiation, and the window maintains a physical barrier between the microscope environment and the detector crystal. The crystal itself converts each X ray into a voltage pulse that is amplified by the field-effect transistor (FET).

The window, located next to the collimator, maintains a physical barrier between the environment of the column of the electron microscope and the environment of the detector crystal. The window protects the detector crystal from vapors that would condense on the cold surface of the crystal, thereby reducing its ability to detect X rays. Most windows are made of a thin sheet of beryllium approximately 7 μm thick.

A major disadvantage of the beryllium window is that it absorbs X rays of approximately 1 keV in energy and less, thereby limiting the ability to detect elements with low atomic numbers, which emit only low-energy X rays. For instance, approximately 44% of the sodium K_α X rays of 1.04 keV are absorbed, and approximately 99% of the oxygen K_α X rays of 0.52 keV are absorbed. The lower detection limit of beryllium window detectors is usually considered to be sodium, atomic number 11.

Various plastic windows and diamond thin-film windows are available and have the advantage of far less absorption of low-energy X rays than that of beryllium windows; typically the lower limit of analysis is atomic number 5, boron. Some of the alternative windows have disadvantages; some types are very fragile, and variations in vacuum level in the microscope may cause them to break. Another potential disadvantage is detector overload with samples that contain large amounts of carbon, nitrogen, and oxygen, such as most biological samples and polymers. Nearly all the X rays from these elements of lower atomic number are detected. The X-ray system can process only a limited number of X rays at a given time; therefore, detection of elements other than those just mentioned is more difficult. Unless there is a specific need for detection of the elements of lower atomic number, the plastic window detector may not be desirable.

Some detectors are designed to be used without a window for short periods of time. A valve, mounted in front of the detector crystal, can be opened when the analysis is to begin. Other detectors have a turret with three positions—one with a beryllium window, one with a plastic window, and one windowless—each selected when required. Windowless operation should be as brief as possible to avoid contamination, unless the detector is mounted on an electron microscope with an ultraclean vacuum system.

The detector crystal converts each X ray into a brief voltage pulse. The crystal consists of a silicon wafer 3.5 mm to 6 mm in diameter and 3 mm thick, into which lithium has been drifted to produce a semiconductor region. When an X ray enters this region, a free electron and a hole are created. A bias voltage of approximately 1,000 V is placed across the crystal to maintain an electron–hole depletion region and to attract the free electron and hole produced by each X-ray pulse. Each 3.8 eV of energy (average) from the X ray is converted into one electron–hole pair. For example, a nickel K_α X ray of 7.471 keV would produce approximately 1,966 electron–hole pairs. The charge resulting from the entrance of an X ray is added to the bias in the form of a brief pulse, the voltage or height of which is proportional to the energy of the X ray.

Energy other than X rays can create electron–hole pairs; backscattered electrons are especially capable, as is radiation. Radioactive energization of the semiconductor can be used to advantage as a standard in testing detector resolution. For instance, ^{55}Fe of known intensity can be placed at a measured distance from a detector; the radiation mimics manganese K_α X rays.

Detector crystals are available with active areas of either 10 mm^2 or 30 mm^2. The sensitivity of the crystal to X rays is directly proportional to the area. The spectral resolution or selectivity of the detector crystals with larger areas is less; typically it is 10% less selective than the 10-mm^2 crystal.

The detector crystal must be kept at liquid nitrogen temperature to prevent redistribution of the lithium, to reduce electronic noise from the crystal, and to prevent the crystal from shorting out from the bias voltage because of the decreased electrical resistance at higher temperatures. The largest component of the detector, the liquid nitrogen dewar, must be refilled once or twice a week.

The field-effect transistor (FET), the first stage of amplification of the voltage pulse from the detector crystal, is located just behind the detector crystal in the same metal tube and, like the crystal, is kept at liquid nitrogen temperatures to reduce electronic noise. The FET separates the voltage pulse from the bias voltage. The output voltage of the FET is a ramp (Figure 8.5, top), which must be reset periodically to zero voltage by an internal electronic circuit.

CONSTRUCTION OF THE EDS X-RAY ANALYZER

The remaining components of the EDS system—the pulse processor, the analog-to-digital converter, the multichannel analyzer, the computer, and the display—are located in separate cabinets some distance from the detector (see Figure 8.3).

The pulse processor changes the ramp output from the preamp into separate voltage pulses, the heights of which are proportional to the energy of the X rays that produced them (Figure 8.5, center). When the pulses arrive too close to each other because of a large number of incoming X rays, the pulse processor shuts off for a brief period of time. This is one cause of what is called dead time, an important component of EDS analysis.

The analog-to-digital converter (ADC) takes each pulse from the pulse processor and converts it into a series of pulses of equal height, the number of which corresponds to the height of the original pulse and

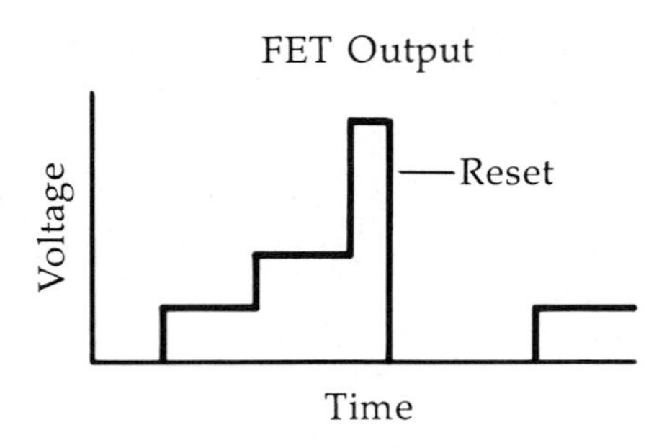

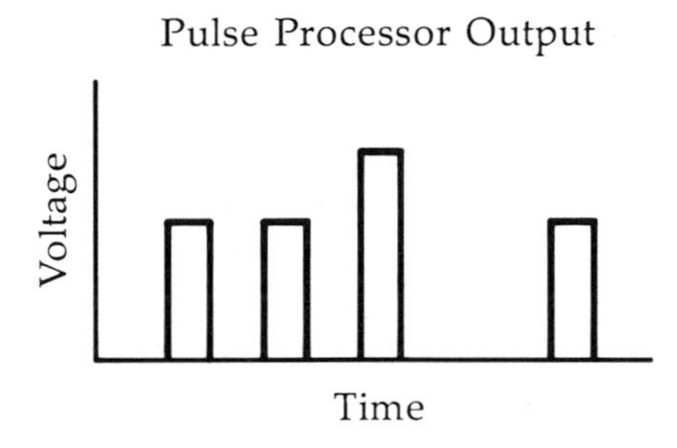

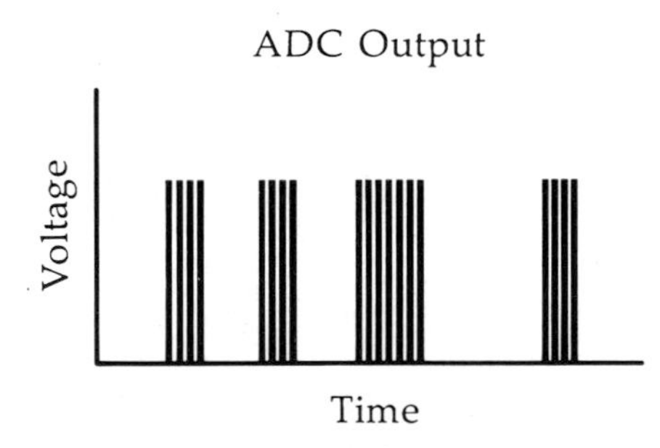

Figure 8.5 Functions of the EDS detector. The output of the field-effect transistor (FET) is a ramp that is periodically reset to zero voltage. The voltage increase on each step of the ramp is proportional to the energy of the incoming X ray. The output of the pulse processor is a series of pulses. The voltage of each pulse is proportional to the energy of the incoming X ray. The output of the analog-to-digital converter (ADC) consists of groups of pulses of similar voltage. The number of pulses in each group is proportional to the energy of the incoming X ray.

to the energy of the X ray (Figure 8.5, bottom). The conversion from analog to digital is necessary so that the complex analysis of the entire range of X rays collected can be done digitally with a computer.

The multichannel analyzer (MCA) takes the digital pulses from the ADC and sorts them into different channels of varying energy. Typically, either 512 or 1,024 channels are used; the energy width of each channel may be 10 eV, 20 eV, 40 eV, etc., depending on the range of energies selected. For instance, an analysis might be done from zero to 10.24 keV using 1,024 energy channels; each channel would then be 10 eV wide. Sodium K_α X rays of 1.04 keV would fall within the energy channel of 1 keV to 1.1 keV. The MCA can count all X rays that fall within that channel.

The computer assists the MCA in performing the counting and sorting of the digital pulses and in addition assists with other functions associated with the X-ray analyzer, such as storing spectra and processing computer programs.

The display is a cathode-ray tube (CRT); most systems use a full-color monitor. The energy spectrum is displayed on the CRT, which is also used for readout of various computer programs associated with EDS analysis.

OUTPUTS

Various outputs or types of data may be obtained from EDS equipment. Qualitative outputs, such as a listing of the elements present (a plot of the spectrum), may be obtained, or the face of the CRT may be photographed showing the spectrum with identified peaks (Figure 8.6A).

Another type of output is called an ROI (region of interest), or a window. An ROI is an area of the spectrum that may be placed around a peak associated with a certain element. The equipment sums the total number of X rays in all the channels within the region. It also establishes a background line between the energy limits and subtracts the total counts beneath this line as background. The numbers from this output can be useful for comparisons.

The spectrum obtained from a scan of the sample yields information about the elements present in the total area scanned; no information is given concerning concentrations of the elements at specific locations. Dot maps show relative concentrations. A dot is placed on the photograph every time an X ray is detected within a preestablished ROI. Areas with greater numbers of dots indicate greater concentrations of the element. Correlating an SEM or STEM image with the dot map reveals which structures contain greater amounts of the element (Figure 8.6B–D).

SPECTRUM ACCUMULATION AND INTERPRETATION

Spectrum accumulation and interpretation are important aspects of EDS analysis. Operation of an EDS system involves many variables, all of which must be optimized to obtain the most accurate data.

Selection of the X-ray energy accumulation range is important. Initially, the X rays should be accumulated in the range of zero keV to 20 keV. All of the elements have useful X-ray lines in this range. In many

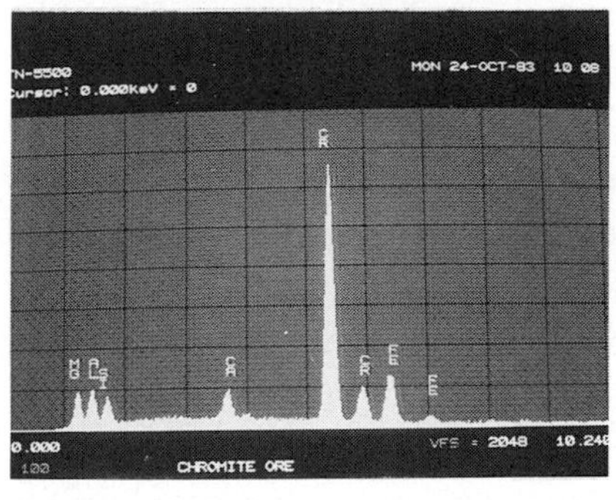

A

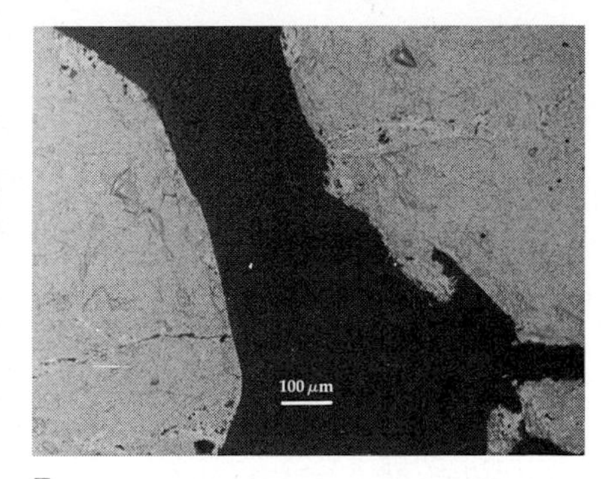

B

C

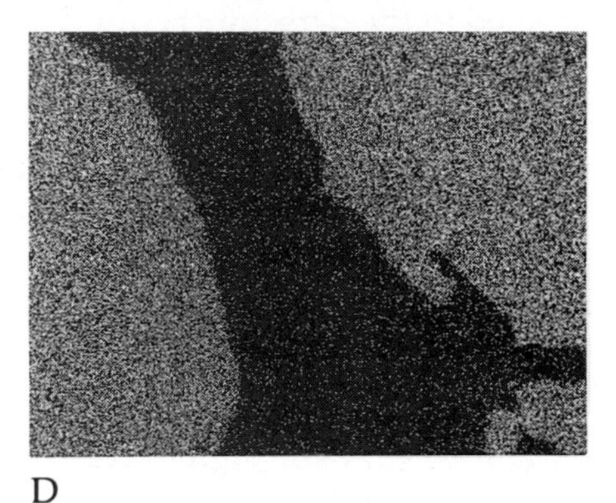

D

Figure 8.6 EDS and SEM analysis of chromite ore. (A) An EDS spectrum of chromite ore showing peaks for magnesium, aluminum, silicon, calcium, chromium, and iron. (B) Backscattered-electron micrograph of the chromite ore. This image provides compositional data based on atomic number. (C) A dot map for silicon (low atomic number) that correlates with the dark areas of the backscattered-electron image. (D) A dot map of chromium (high atomic number) that correlates with the light areas of the backscattered-electron image.

instances, an element has several lines in this range, facilitating accurate identification.

Selecting the proper accumulation time is important. Accumulation times in the range of 30 seconds to 10 minutes are common. Longer accumulation times, in general, give more accurate data, because they produce an X-ray peak more clearly above the background level and provide more data for analysis. Detecting trace elements usually requires more time.

The beam current should be adjusted to an acceptable range for proper data collection. The greater the beam current, the more X rays produced. It is not always desirable, however, to use the maximum amount of beam current. The EDS pulse processor can process only a limited number of X rays per second; usually the maximum rate is about 10,000 counts per second. The beam current affects the dead time in the pulse processor. A dead time of 5% to 30% is desirable; below this range few X rays are being detected, and above this range artifacts form. The beam current can be measured. In an SEM, a device called a Faraday cup intercepts and traps the beam of electrons to allow measurement of the current. In a TEM, the current induced in an aperture by scattered electrons (called aperture current) is used as an indication of beam current. Both of these methods allow the beam current to be adjusted to some standard value, which is helpful when comparisons are needed.

The end result of the proper selection of energy range, accumulation time, and beam current is a spectrum of energy displayed on the CRT. A spectrum consists of a series of peaks displayed on the background. Each peak will have a Gaussian shape due to the variation of the actual X rays and the variable number of the electron–hole pairs created in the detector crystal. The first step in identifying the peaks is to match the locations of the lines of the various elements with the peaks. The locations of the lines should match the centers of the peaks well. The second step requires matching the relative intensities of the lines (called weights) with the relative intensities of the peaks. The weights of the lines are as follows: $K_\alpha = 1$, $K_\beta = 0.1$, $L_\alpha = 1$, and $L_\beta = 0.7$.

These relationships imply that if a strong peak is associated with a K_α line, there should be a peak associated with the K_β line that is approximately one-tenth as intense. Likewise, if there is a peak thought to be from an L_α line, there should be a peak associated with the L_β line that is seven-tenths as intense. The α peaks are always more intense than the β peaks because it is more probable that one jump will occur than a jump of two lines. The β peaks are always slightly higher in energy than the α peaks because of a greater energy difference between the shells. The relationship of α to β holds true only for a given family of lines (K, L, etc.). No direct comparisons between intensities of K-line peaks and L-line peaks should be made for an element because these are affected by many variables. Normally if a K_α peak is present, there will be a K_β peak, and if an L_α peak is present, there will be an L_β peak, unless the α peaks are weak, in which case the β peaks may not be detectable.

Sometimes only one line matches with peaks, and sometimes several do, depending on the element and the energy range. It may be possible to match two families of lines; seldom is it possible to match three families. The K lines always have the highest energy level, L lines are intermediate in energy, and M lines have the lowest energy level. With gold, for example, the K_α line is at 68.185 keV, the L_α line is at 9.711 keV, and the M_α line is at 2.123 keV.

Occasionally a spectrum contains areas in which it is difficult to determine if a peak is present; random distribution of background X rays can produce areas that resemble a peak. Although statistical analysis of X-ray data can be complex, a widely accepted rule is that if the net number of counts above the background is three times or more the square root of the number of background counts under the suspected peak, there is a 99% chance that the suspected peak is real.

OPTIMIZING THE DETECTION OF X RAYS

The detector can be moved toward the beam either manually by a crank or with an electric motor. The X distance between the detector and the beam affects variables called takeoff angle and solid angle, which in turn affect the number of X rays detected.

Takeoff angle is the angle between the surface of the specimen and a line drawn from the center of the detector to the location of the beam on the sample. X-ray detection is by line of sight. X rays move in straight lines, and only those moving in a direction toward the detector are detected. Therefore, the X rays detected in a position with a large takeoff angle will have gone through less of the sample than those detected in a position with a small takeoff angle (Figure 8.7). Since the sample absorbs X rays, more X rays are detected in the position with the large takeoff angle. Absorption of the X rays by the sample due to takeoff angle is of more concern with whole samples than it is with thin sections because in thin sections usually not enough sample is present for significant absorption of X rays. The takeoff angle can be increased in

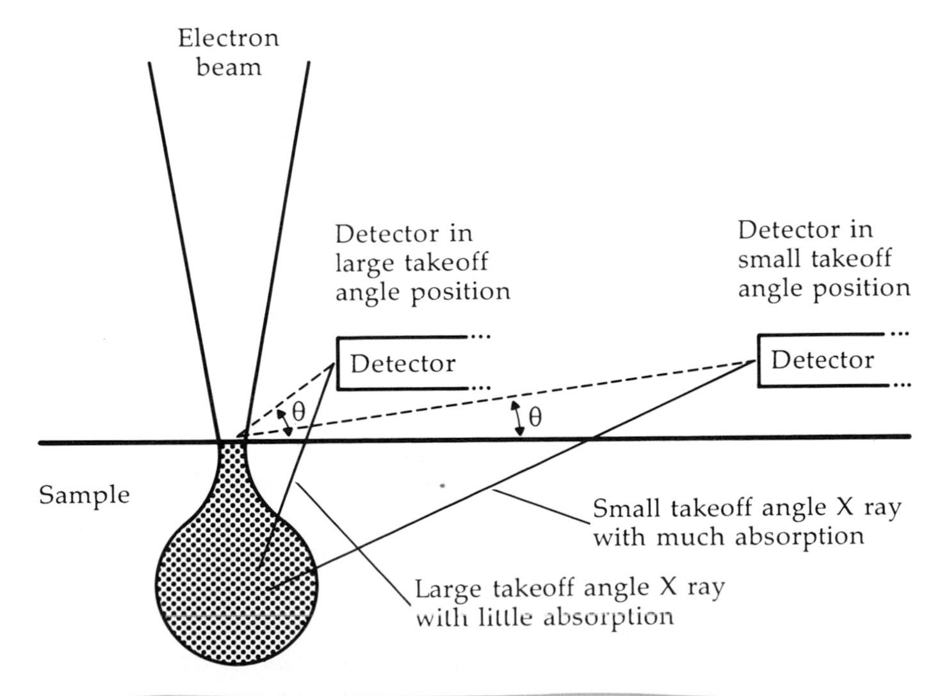

Figure 8.7 Detector distance and takeoff angle. X rays collected from a large takeoff angle have undergone less absorption than those collected from a small takeoff angle.

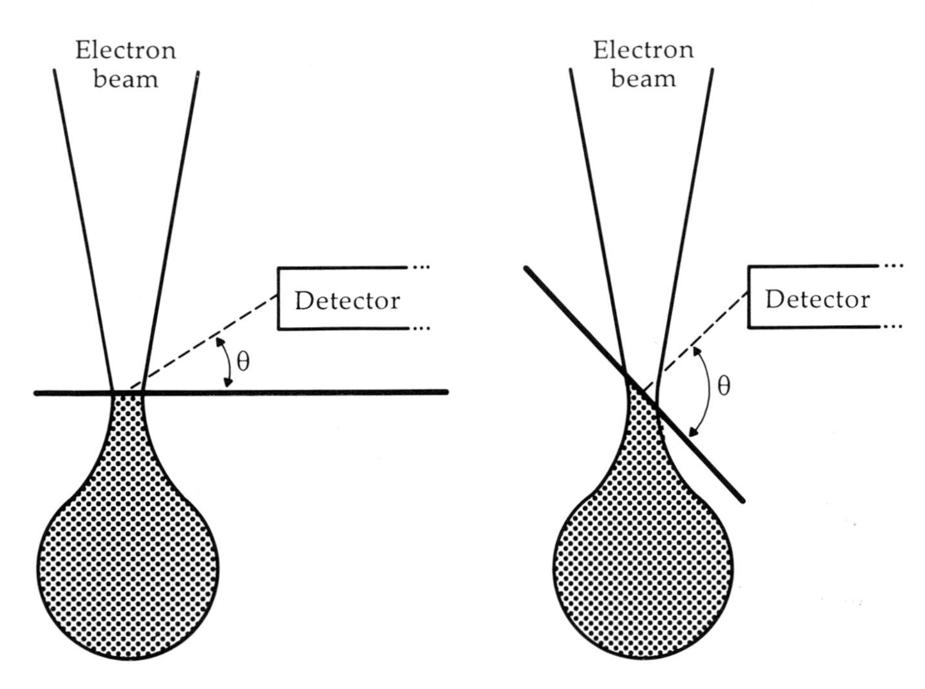

Figure 8.8 Sample tilt and takeoff angle. A tilted sample produces a larger takeoff angle and therefore less absorption of X rays.

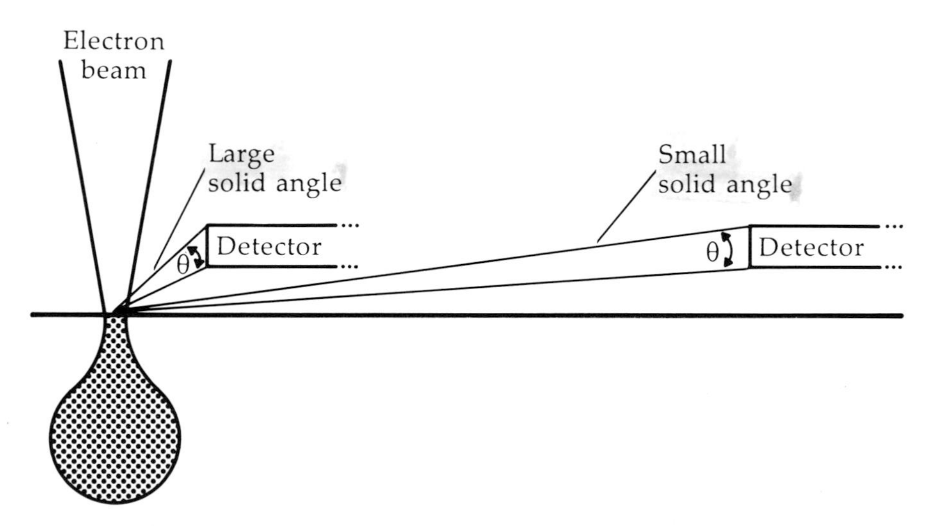

Figure 8.9 Detector distance and solid angle. At short distances, the detector collects a higher percentage of X rays from the solid angle.

one of two ways: by moving the detector closer to the beam (see Figure 8.7) or by tilting the sample (Figure 8.8).

Solid angle is the angle subtended by the detector (Figure 8.9). A large solid angle results in more X rays being detected than a small solid angle does. Moving the detector closer to the beam increases the solid angle, except at long working distances often encountered in SEMs (Figure 8.10), where a zero solid angle will result if the detector is too

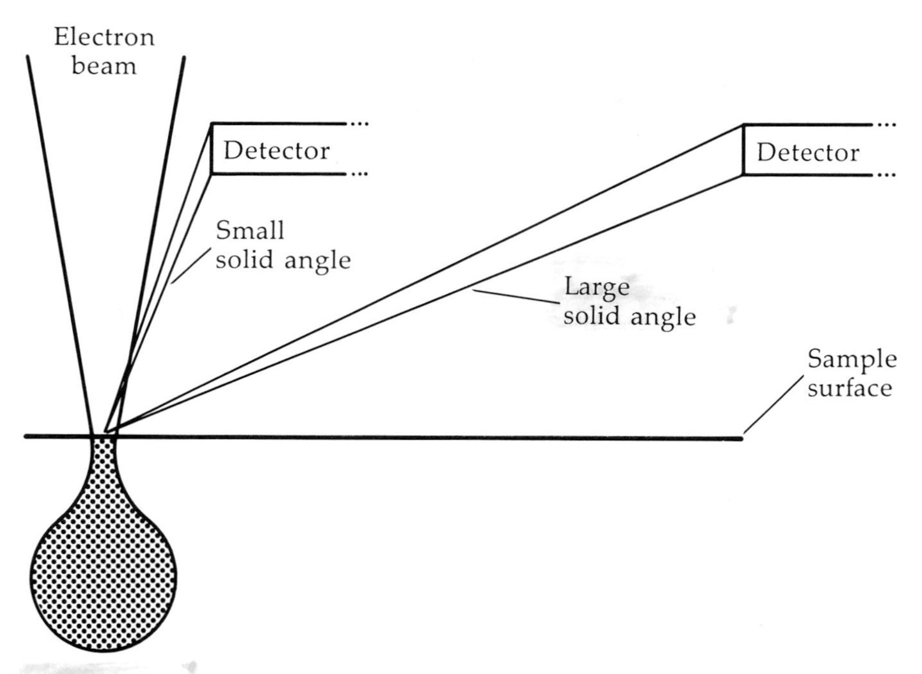

Figure 8.10 Detector distance and solid angle at long working distances. At long working distances, a larger solid angle of collection will be obtained with the detector located at a distance from the sample.

close to the beam. At long working distances, an intermediate detector position produces the greatest solid angle.

Solid angle and takeoff angle are interrelated: Moving the detector closer to the beam can result in more X rays being detected because of an increase in both the solid angle and the takeoff angle.

ARTIFACTS

EDS is prone to a number of artifacts. It is essential that these artifacts be recognized and minimized.

System Peaks

Ideally, only the area of the sample under electron bombardment produces X rays. In practice, this may not be true. Instead, X rays are produced from areas of the sample, from the sample support and holder, and from other components of the microscope sample chamber that are not under electron bombardment. The X rays from these other areas produce the artifacts known as system peaks.

The final aperture in an SEM, especially the thin-foil type, can be a major cause of system peaks. Scattered electrons that impinge

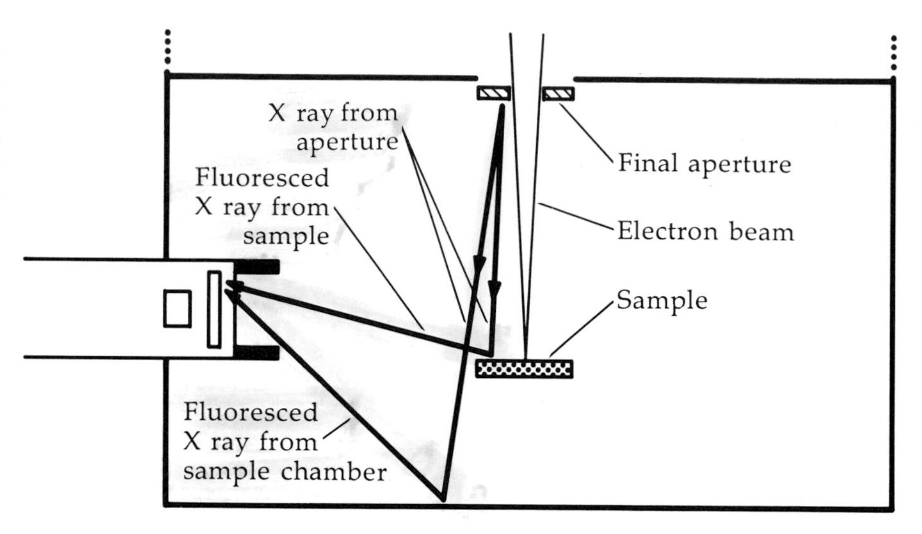

Figure 8.11 System peaks from final aperture fluorescence. The electron beam produces X rays from a thin aperture that in turn fluoresces X rays from various parts in the sample chamber. These X rays are artifacts and are difficult for the detector to separate from the X rays produced by the specimen–beam interactions.

away from the aperture opening are capable of producing X rays within the aperture material. Because this material is thin, the X rays can exit on the sample side, continue downward and strike the sample, the stub, the stub holder, and so on (Figure 8.11). X rays are capable of producing more X rays when they strike any material in much the same manner that electrons can, through a process called fluorescence. The detector has very limited capability of separating these X-ray artifacts from the X rays produced by the beam; the capability that it does have is due to the collimator. System peaks derived from final-aperture fluorescence can be minimized by using a hard X-ray aperture, usually a thin-foil molybdenum aperture that has been coated with tungsten around the aperture opening to absorb the X rays produced by the aperture material. In addition, coating the sample holders with carbon cement and using graphite stubs can reduce these peaks greatly.

Aperture-induced system peaks occur in the TEM in a similar way as in the SEM. These peaks often are derived from the anticontaminator, the sample rod, and other parts of the system. Hard X-ray apertures made of carbon are available to reduce such peaks. In addition, carbon or nylon grids and carbon sample holders can be used to reduce system peaks.

Another source of system peaks of concern mainly in the SEM is backscattered electrons from the sample, which may strike other components in the sample chamber, such as the walls or lens pole pieces and produce X rays (Figure 8.12). Minimizing the production of this type of system peak can be difficult; materials near the sample, especially those directly above, may be moved, coated with carbon cement, or lined with graphite.

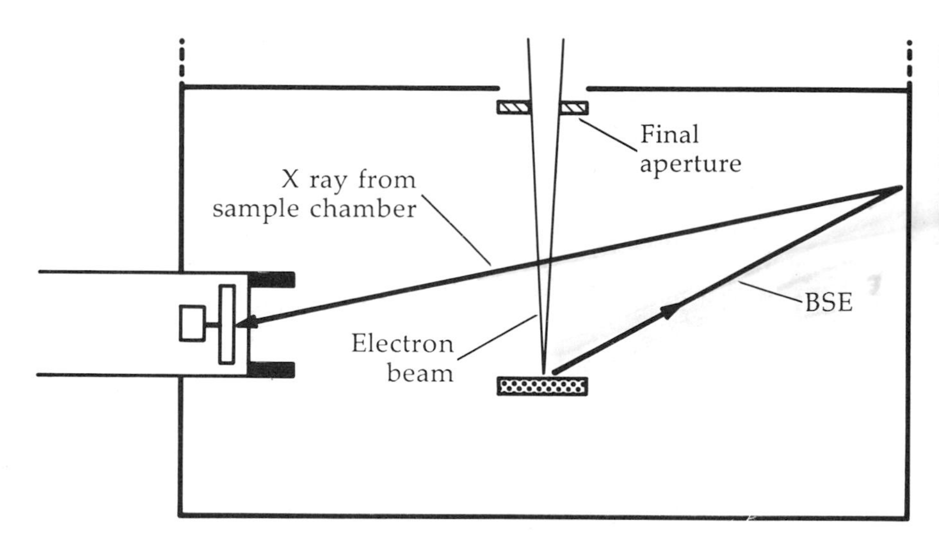

Figure 8.12 Backscattered electrons and system peaks. Backscattered electrons (BSE) from the sample may strike various parts of the sample chamber and produce X rays. These X rays often are difficult to minimize and can cause artifacts in the data.

Sum Peaks

Sum peaks occur when two X rays arrive either simultaneously, or nearly at the same time, as a result of the inability of the pulse processor to identify them as two distinct X rays. The energies of the two X rays are then combined, creating a false peak at their sum. Sum peaks are seldom a problem unless the EDS system is malfunctioning or is being operated improperly, such as with a very high dead time. If a sum peak is suspected, a new accumulation should be done with a lower dead time. If the ratio of the suspected sum peak to other peaks changes substantially, then the probability is very high that the former was a sum peak.

Escape Peaks

When X rays strike the detector crystal, instead of producing electron–hole pairs, they may produce silicon X rays, most of which will be absorbed. The silicon X rays that are not absorbed will escape from the crystal and subtract the silicon K_α energy (1.74 keV) from the energy being measured, creating a false peak (called an escape peak) at –1.74 keV from the main peak.

Escape peaks can be a very serious problem if not recognized. The production of escape peaks is a process of fluorescence because an X ray is producing another X ray. X rays with energy just above the silicon K critical excitation energy of 1.838 keV are very efficient at producing escape peaks. For instance, phosphorus K_α X rays at 2.015 keV and silver L_α X rays at 2.984 keV produce escape peaks of about 1% to 2% of the magnitude of the main peak, whereas zinc with a K_α peak at 8.638 keV produces an escape peak at less than 0.01% of the magnitude of the main peak. X rays with energy less than the critical excitation energy of silicon

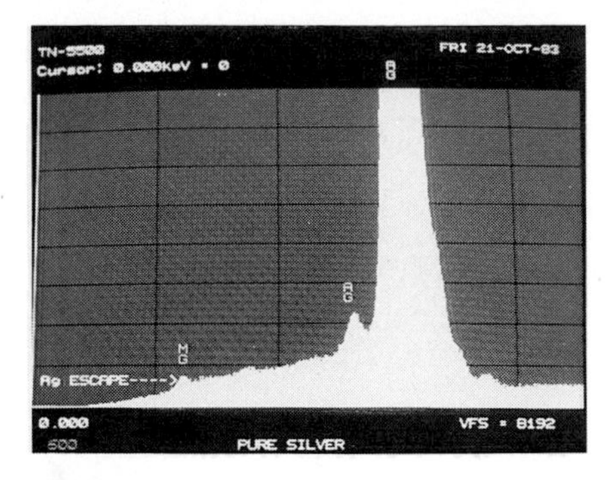

Figure 8.13 Escape peaks. A spectrum from pure silver. The escape peak shown coincides with the K line for magnesium (Mg) and could be mistaken for that element.

at 1.838 keV do not produce escape peaks. Whenever a spectrum is obtained with large peaks in the range of 1.74 keV to 4 keV, escape peaks should be suspected and their locations marked to ensure that they are not mislabeled. For instance, a spectrum with a strong silver L_α peak at 2.984 keV produces an escape peak at 1.244 keV, which could be misidentified as magnesium with a K_α energy of 1.254 keV (Figure 8.13).

Secondary Fluorescence

Secondary fluorescence refers to the process of fluorescence (i.e., X rays inducing X rays in a material) within the sample. X rays from a given element in the sample may produce X rays from a second element at a different location if the second element has an critical excitation energy that is less than the characteristic energy of the first element. For example, with the beam on an accumulation of chlorine, some of the chlorine K_α X rays at 2.622 keV strike the phosphorus concentration and produce phosphorus K_α X rays because the critical excitation energy of phosphorus is 2.142 keV (Figure 8.14). The phosphorus thus appears to be at the same location as the chlorine. Not only X rays are capable of leading to errors of this type; backscattered electrons are capable of producing a similar effect, although technically it is not called secondary fluorescence.

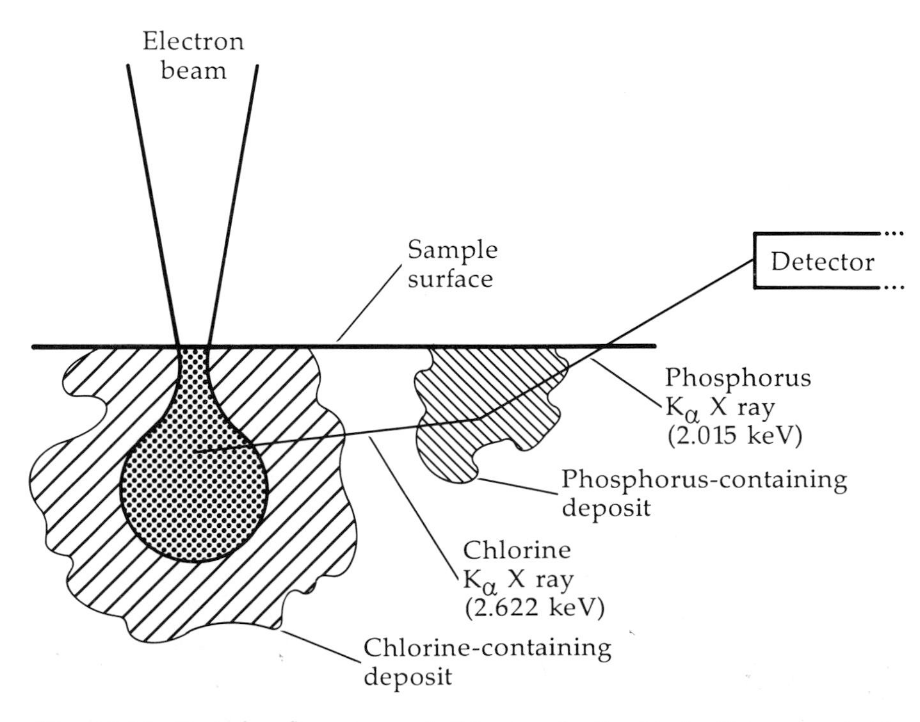

Figure 8.14 Secondary fluorescence. In this example, the chlorine X rays are capable of causing fluorescence of phosphorus X rays. The phosphorus would erroneously appear to be at the same location as the chlorine.

Specimen Topography

The topography of the sample can produce a variety of artifacts in the X-ray data obtained. Ideally, all samples should be perfectly flat and smooth, as most thin sections are. Whole samples used in the SEM often are not flat. The topographical artifacts arise from absorption and blockage of X rays. Absorption occurs from local variations in the takeoff angle on an irregular sample surface. The X-ray peaks obtained from two points in a homogenous sample can differ considerably in height because of sample absorption. If the sample has extreme irregularities, the absorption may be complete, causing blockage and detection of no X rays.

Mass Loss

Mass loss is loss of material from the sample as a result of interaction with the electron beam. Biological material, especially in thin section, is prone to mass loss, as are halogens in crystalline form, such as sodium chloride (NaCl) and potassium chloride (KCl). Losses of up to 30% in a normal scan of several minutes are common. Mass loss may result in reduced counts if an element of interest is being lost; conversely, it may result in increased counts for a certain element if other elements are being lost, which effectively increase the concentration of the element being retained. The background X rays may be monitored as an indication of the rate of mass loss. A cold stage that can maintain sample temperatures of –130°C or less effectively eliminates mass loss.

QUANTITATIVE ANALYSIS

Quantitative analysis of the elements in a sample is a complex task that requires sophisticated computers and software. The methods used for whole samples differ drastically from those used for thin sections. Certain portions of the software quantitation programs (those for measuring the peak intensity and those for peak deconvolution) are often similar.

Peak-Intensity Measurement and Peak Deconvolution

Measuring peak intensity by removing background counts is crucial for accurate quantitative analysis. Two common methods for measuring peak intensity are background modeling and background filtering. Both are complex computer-assisted procedures relying on theoretical mathematical descriptions of the X-ray spectra. The software for the method is usually included as part of the quantitation program available from the EDS equipment manufacturer. One program in

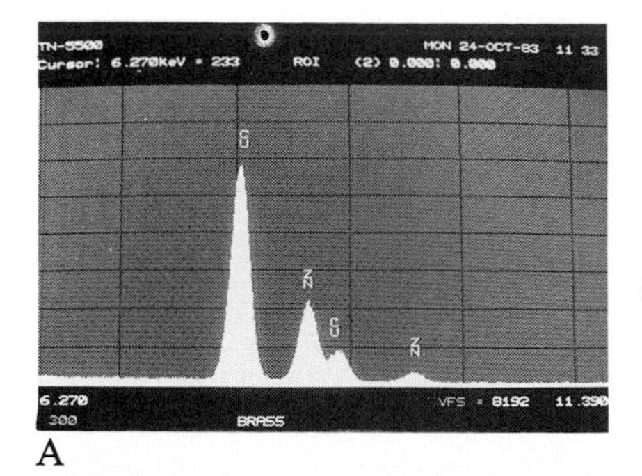

A

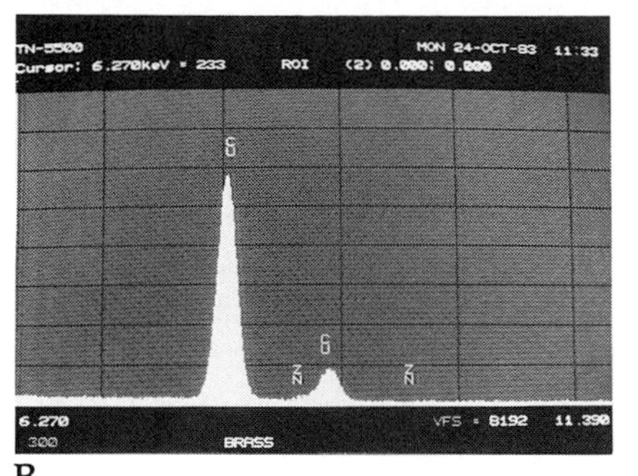

B

Figure 8.15 Deconvolution of peak overlaps. (A) A spectrum of brass showing overlap of the copper K_β peak with the zinc K peak. (B) The same spectrum after the computer deconvoluted the peaks and removed the zinc from the spectrum.

particular, the "top hat" digital filtering version of background filtering, has gained widespread acceptance.

Peak deconvolution is often necessary to remove the overlaps encountered in EDS. For accurate quantitative analysis, overlaps must be removed so that X-ray intensities can be assigned to the proper elements. Procedures for removing overlaps have various names, such as peak deconvolution, peak stripping, or curve fitting; all are complex computer programs that are normally included as part of a software package sold by the EDS equipment manufacturer. One of the most popular methods, called super multiple linear least squares, requires standards for the elements being fitted and deconvoluted. If the proper criteria are satisfied, the deconvolution procedures are capable of extremely accurate results (Figure 8.15).

ZAF Quantitation of Whole Samples in the SEM

Ideally, the number of X-ray counts for an element in an unknown substance could be compared to the number of X-ray counts for the element in a pure standard, and the ratio would be equal to the concentration. For instance, in an alloy of 92% copper and 8% aluminum, the aluminum would produce 8% of the X rays produced by the pure aluminum standard. Unfortunately, it actually produces only about 2% of the number produced by the standard because of complex interactions (called matrix effects) between the aluminum and the copper. The ratio obtained, 0.02 in this instance, is called the K ratio and is only the first approximation of the true concentration. A correction must be applied to the K ratio to compensate for the matrix effects. One of the most common correction routines, the ZAF correction, consists of three separate corrections: Z, A, and F.

The first of these, the Z factor, refers to the effect of the atomic number of the sample compared to the standard. The Z factor is composed of two subfactors, R and S. R compensates for the backscattering effect. In the 92% copper/8% aluminum example, more beam electrons are lost to backscattered-electron production in the sample than in the standard; thus, fewer aluminum X rays would be produced, and R would be greater than one to compensate. The S factor compensates for the stopping-power effect. In the example, the stopping power of the sample is greater than the standard; thus, more aluminum X rays would be produced, and S would be less than one to compensate. The total Z factor in this example is $R\,(>1) \times S\,(<1) = 0.916$. The Z factor varies with accelerating voltage because it affects both the R and the S subfactors.

The A factor refers to differences in X-ray absorption between the sample and the standard. In the example given, aluminum X rays would be absorbed more strongly in the sample than they would be in the standard because of the copper. Therefore, fewer aluminum X rays would be produced, and A would have to be greater than one to compensate. (In the example, A is 4.285.) The A factor is affected by differences in accelerating voltage and takeoff angle and by the other elements in the sample that absorb X rays.

The F factor refers to fluorescence. In the example, some of the copper X rays fluoresce X rays from aluminum. More aluminum X rays are produced than were expected, and F is less than one (0.999) to compensate. The accelerating voltage, takeoff angle, and family of lines being analyzed all affect the F factor. The closer the critical excitation energy to the characteristic energy of another element, the greater the F factor.

The total ZAF correction in the 92% copper/8% aluminum sample would be Z (0.916) × A (4.285) × F (0.999) = 3.925. The K ratio, 0.02, multiplied by 3.925 equals 0.785, which indicates 7.85% aluminum, well within the accepted range of accuracy for EDS quantitation.

The values used for the ZAF factors normally are calculated by computer using the quantitation software supplied by the manufacturer. The calculated values for each factor vary with accelerating voltage, takeoff angle, presence and amount of other elements in the sample, and the X-ray lines being used for analysis; the values might thus be different from sample to sample. It is always necessary, however, to ensure that the same accelerating voltage, takeoff angle, and accumulation time is used for all samples and standards. If any of these three factors are changed, then all the standards and samples have to be scanned again using the new set of conditions.

The standards used in ZAF correction factor analysis are either pure elements or compounds such as minerals. Pure elements are used when they exist in a form that can be placed in an electron microscope, and minerals (or compounds) are used for elements like sodium, phosphorus, chlorine, potassium, and calcium that cannot. Multielement standards are available from a number of suppliers and usually contain several dozen pure elements and compounds mounted firmly and polished smooth to yield extremely accurate results.

The use of minerals as standards requires using ZAF correction factors on the standard; the weight percent will not give a proper correction factor. The situation is somewhat different from a normal unknown sample, because the atomic number percent and weight percent are known or may be calculated from stoichiometry. The ZAF correction may then be run backward to give a K ratio. The K ratio, in this instance, is not a true K ratio because the sample is the standard. The number calculated for the K ratio, however, may be used as a correction factor equivalent to the purity of the element. For instance, sodium chloride (NaCl) may be used as a standard for both sodium and chlorine; it is 39.66% sodium by weight and 60.34% chlorine by weight. It is not 39.66% sodium relative to the number of X rays produced, however, because of the ZAF effects: Z = 0.992, A = 1.716, and F = 0.997, so ZAF = 1.699. Relative to the X rays produced, NaCl is 23.3% sodium (0.3966 ÷ 1.699 = 0.233).

ZAF correction computer programs that do not require the use of external standards are available. With these programs, standards for the elements are entered into memory at the time the software is written. Most of these programs are sophisticated and are capable of giving very accurate results if the design criteria are followed. One major limitation is that no nondetectable elements (such as carbon, nitrogen, or oxygen) can be present if a beryllium window detector is being used. If nondetectable elements are present, the program cannot analyze for their presence, and it will make the total of all detectable elements equal to 100%, when in fact that is not true. In addition, the relative ratios of the elements may not be accurate. If the nondetectable elements are present in the form of oxides of known formulas, however, the formulas can be entered into the program, and it can compensate for their presence.

Non-ZAF Quantitation for Whole Biological Samples in the SEM

The use of ZAF correction with pure elements or minerals for standards has had limited success in biology because of two factors: The sum of detectable elements or their oxides does not add up to 100% (and in most instances is 10% or less), and the concentration of elements with atomic numbers greater than 10 is very low. Both of these factors are considerably different from the design criteria of the ZAF correction procedure.

The ZAF correction factors may not be needed in the analysis of many biological samples; the K ratio is thus used as the final concentration. Four criteria must be met to eliminate the need for correction factors. First, standards similar in composition to the tissue must be used, such as frozen hydrated or freeze-dried gelatin with various salts of known concentration added. Second, the standards should have elemental concentrations similar to those in the tissue being analyzed. Third, the concentration of any single element must be less than 10%. Fourth, there must be no major concentrations of elements with atomic numbers greater than 20; such concentrations would cause fluorescence and would require a correction factor.

Most biological samples meet these four criteria. All of the elements normally of biological interest (sodium, magnesium, phosphorus, sulfur, chlorine, potassium, and calcium) have an atomic number less than 20. The four criteria ensure that any required ZAF corrections would be extremely small and, in most instances, could be neglected.

Bence-Albee Quantitation Method for Geological Samples

A common method of quantitation in geology is called the Bence-Albee procedure. This empirical method compares unknown minerals with standards consisting of minerals of very similar composition.

Although Bence-Albee programs can be used with EDS, they work far better with WDS systems because of the problem of peak overlaps in the standards. Modern EDS ZAF correction programs are almost as accurate as the Bence-Albee procedure.

Quantitation of Thin Biological Sections in the TEM

Quantitative measurement of elemental concentrations in thin samples requires methods different from those used for thick or bulk samples. By definition, thin samples are those through which the incident electron beam passes with little loss of energy and from which the resulting X rays escape with little absorption. Several methods exist for quantitation of thin samples. The most common is the Hall method.

The Hall method presumes the use of thin sections in the range of 100 nm to 500 nm thick, with no large concentrations of elements, especially those with atomic numbers greater than 20; thus, ZAF corrections are not needed. The quantity of X rays from the element being measured is proportional to the number of atoms of that element in the area of specimen–beam interaction; therefore, the quantity of X rays is also proportional to section thickness, and slight variations in section thickness cause variations in the quantity of X rays. The quantity of background X rays is proportional to the total number of atoms of all elements in the area of specimen–beam interaction. The peak-to-background ratio is a measure of concentration, because it relates the number of atoms of the element being measured to the total number of atoms present. In addition, the peak-to-background ratio compensates for variations in section thickness because if a section is slightly thicker, the background increases proportionately with the peak, and the ratio is constant. For the same reason, the peak-to-background ratio compensates for variations in beam current.

The standards used for the Hall quantitation method usually consist of a salt dissolved in resin, which is sectioned using conventional techniques, or a salt dissolved in gelatin or albumin, which is sectioned using cryogenic techniques. In summary, the Hall procedure uses a net peak-to-background ratio for an element in a section as an indication of the concentration of that element. The background can be measured in any part of the spectrum that is free from characteristic peaks. It is important that the background be true background from the specimen; the contribution from the support film, grid, and system background must be measured and subtracted from the measured background.

Quantitation of Metallurgical Thin Films

Quantitative analysis of metallurgical thin films usually is done using the Cliff-Lorimer procedure. The procedure relates the ratio of the concentrations of two elements (A and B) to the measured ratio of the X-ray intensities for each multiplied by a correction factor (K_{ab}) called the proportionality factor, or the Cliff-Lorimer factor. This factor is usually determined using known standards, although it may be calculated. For three or more elements, A may be calculated with reference to B, B may be calculated with reference to C, and so on for all elements so that the total equals 100%. The program usually assumes that sample absorption and fluorescence are minimal and may be ignored. In samples where absorption may be significant, the foil thickness is measured and used to calculate an absorption correction factor.

SAMPLE PREPARATION

Sample preparation for X-ray analysis is more exacting than that for conventional electron microscopy. Great care must be taken to ensure that the preparation procedure does not remove elements of interest or add elements that are not normally present. Samples for quantitative analysis usually must be flat and smooth, and nonconductive samples must be made conductive without using a heavy metal.

Preparation of Whole Materials-Science Samples for SEM Analysis

Materials-science samples such as metals, polymers, ceramics, and minerals, can be prepared for qualitative X-ray analysis using normal SEM sample preparation procedures. If quantitative X-ray analysis is to be done, much more exacting requirements must be met. In most circumstances the sample must be flat and smooth to avoid the geometric effects discussed earlier. Standard geological or metallurgical cutting and polishing methods are usually suitable. (See Further Reading for more information.)

Preparation of Whole Biological Samples for SEM Analysis

Conventional chemical SEM sample preparation procedures can be used only in situations in which the elements of interest are either firmly bound or are insoluble (such as plant crystals, kidney precipitates, silicon in plants, and calcifying systems such as teeth and bones). Although conventional techniques can be used in other situations than those described, the loss of elements may range from about 10% to 95%. Addition of elements from fixatives and buffers is very likely with conventional techniques.

Freeze-drying the sample works well in many situations. The sample should be frozen quickly using rapid freezing methods or plunged into liquid nitrogen slush to minimize soluble element relocation or ice crystal damage. Movement of the elements that exist in the form of soluble salts will occur. The salts, however, are generally believed to adhere to the nearest unit membrane. In most instances the distance of element redistribution to the nearest unit membrane would be less than the analytical spatial resolution for a bulk sample (up to 30 μm).

Air-drying of a sample with no fixation or dehydration gives excellent results for certain samples with bound elements within a rigid structure, such as silicon in the plant *Equisetum* sp. and calcium in teeth and bones.

Maintaining samples in the frozen-hydrated state is one of the best sample preparation methods for X-ray analysis because elements normally do not migrate significantly, as they do with air-drying or freeze-drying. For some studies, such as elemental analysis of fluids in vacuoles or extracellular space, this technique is the only one that works. In addition, any measured elemental concentrations are realistic because they are based on values with the normal cellular fluids present. The presence of water, however, can be a major disadvantage compared to freeze-drying because it reduces the percent concentration (by weight) of the elements by a factor of approximately ten, thus making detection of the elements more difficult.

Preparation of Thin Materials-Science Samples

A variety of methods can be used to produce a thin sample from a whole sample, including ion beam thinning and dimpling. (See Further Reading for more information.)

Preparation of Thin Biological Sections

Conventional preparation methods (fixation, dehydration, embedding, and sectioning) can lead to losses of up to 90% for some elements or to the addition of elements not normally present. Conventional methods can be used in special situations, such as natural inclusion bodies, deliberately added compounds (such as drugs), histochemical reaction products, or where elements are firmly bound, as in chromosomes.

Freeze-substitution (see Chapter 6) can be used to prepare samples for X-ray analysis. For X-ray analysis, the method consists of five steps: (1) quick-freezing the sample to minimize ice crystal damage, (2) substituting the ice with a mixture of diethyl ether and acrolein at −90°C for several days, (3) gradually warming the sample to room temperature over a period of several days, (4) infiltrating the sample with epoxy or modified acrylic resin, and (5) sectioning the sample. Sections in the range of 0.25 μm to 1 μm thick are usually used; thus, STEM imaging is required. Freeze-substitution, when compared to cryosectioning, has the advantages of easier sectioning and the ability for sample storage. It has been criticized for elemental loss and elemental relocation.

Cryosectioning is an excellent sample preparation technique, although it is considerably more difficult than standard sectioning. The method consists of quickly freezing the sample and then sectioning it in the frozen state to produce either thick or thin sections that are examined in the frozen state or are freeze-dried and then examined. Efficient, rapid freezing is absolutely necessary to ensure full use of the analytical spatial resolution capabilities of the thin to ultrathin sections (see Chapter 6). Following freezing, the tissue is sectioned in a cryo-ultramicrotome. The sections are then freeze-dried in the microscope column or vacuum evaporator and examined at room temperature or are examined in the frozen-hydrated state with an accessory sample holder for the TEM or STEM that maintains the sample at temperatures of about –150°C.

Carbon Coating for Conductivity

Whole, nonconductive samples examined in a SEM or thick nonconductive sections examined in a STEM usually require a conductive coating. Conventional coatings of metals absorb a high percentage of the X rays produced and add strong X-ray peaks to the spectrum, making analysis difficult. Normally, samples requiring a conductive coating are coated with a thin coating of carbon using either a string evaporator (see Chapter 7) or a vacuum evaporator (see Chapter 6).

FURTHER READING

Boekestein, A., A. L. H. Stols, and A. M. Stadhouders. 1980. Quantitation in X-ray microanalysis of biological bulk specimens. *Scanning Electron Microsc.* 1980(2):321–334. (Whole biological sample quantitation)

Echlin, P., and A. J. Saubermann. 1977. Preparation of biological specimens for X-ray microanalysis. *Scanning Electron Microsc.* 1977(1):621–637. (Analytical resolution, biological sample preparation)

Edelmann, L. 1986. Freeze-dried embedded specimens for biological microanalysis. *Scanning Electron Microsc.* 1986(4):1337–1356. (Freeze-dried sections)

Erasmus, D. A. 1974. The application of X-ray microanalysis in the transmission electron microscope to a study of drug distribution in the parasite *Schistosoma mansoni* (Platyhelminthes). *J. Microsc.* 102:59. (EDS in drug distribution)

Gabriel, B. L. 1985. *SEM: A User's Manual For Materials Science.* American Society for Metals, Metals Park, OH. (General EDS, polishing of materials-science samples)

Geller, J. D. 1977. A comparison of minimum detection limits using energy and wavelength dispersive spectrometers. *Scanning Electron Microsc.* 1977(1):281–288. (Comparison of EDS and WDS)

Goldstein, J. I., D. E. Newbury, P. Echlin, D. C. Joy, C. Fiori, and E. Lifshin. 1981. *Scanning Electron Microscopy and X-Ray Microanalysis. A Text for Biologists, Materials Scientists, and Geologists.* Plenum, New York. (Instrumentation, artifacts, quantitation)

Goldstein, J. I., D. B. Williams, and G. Cliff. 1986. Quantitative X-ray analysis. In *Principles of Analytical Electron Microscopy*. D. C. Joy, A. D. Romig, Jr., and J. I. Goldstein, eds. Plenum, New York. (Cliff-Lorimer quantitation)

Goodhew, P. J. 1973. *Specimen Preparation for Materials Science*. American Elsevier, New York. (Preparation of thin materials-science samples)

Hall, T. A. 1979. Biological X-ray microanalysis. *J. Microsc.* 117:145–163. (Thin-section quantitation of biological samples)

Hall, T. A., H. C. Anderson, and T. Appleton. 1973. The use of thin specimens for X-ray microanalysis in biology. *J. Microsc.* 99:177–182. (Thin-section quantitation of biological samples)

Johnson, D. E., and M. E. Cantino. 1986. High resolution biological X-ray microanalysis of diffusable ions. In *Advanced Techniques in Biological Electron Microscopy*, vol. 3. J. K. Koehler, ed. Springer-Verlag, New York. (Thin-section quantitation of biological samples)

Lever, J. D., R. M. Santer, K. S. Lu, and R. Presley. 1977. Electron probe X-ray microanalysis of small granulated cells in rat sympathetic ganglia after sequential aldehyde and dichromate treatment. *J. Histochem. Cytochem.* 25:295. (EDS in histochemistry)

Marshall, A. T. 1980. Freeze-substitution as a preparation technique for biological X-ray microanalysis. *Scanning Electron Microsc.* 1980(2):395–408. (Freeze-substitution)

Marshall, A. T. 1980. Frozen-hydrated sections. In *X-Ray Microanalysis in Biology*. M. A. Hayat, ed. University Park Press, Baltimore. (Biological sample preparation)

Marshall, A. T. 1980. Principles and instrumentation. In *X-Ray Microanalysis in Biology*. M. A. Hayat, ed. University Park Press, Baltimore. (Instrumentation)

Marshall, A. T. 1980. Quantitative X-ray microanalysis of frozen-hydrated bulk biological specimens. *Scanning Electron Microsc.* 1980(2):335–348. (Biological sample preparation)

Morgan, A. J. 1979. Non-freezing techniques of preparing biological specimens for electron microprobe X-ray microanalysis. *Scanning Electron Microsc.* 1979(2):67–80. (Biological sample preparation and quantitation)

Morgan, A. J. 1980. Preparation of specimens. Changes in chemical integrity. In *X-Ray Microanalysis in Biology*. M. A. Hayat, ed. University Park Press, Baltimore. (Biological sample preparation)

Roomans, G. M. 1979. Standards for X-ray microanalysis of biological specimens. *Scanning Electron Microsc.* 1979(2):649–657. (Biological thin-section quantitation standards)

Roomans, G. M. 1980. Problems in quantitative X-ray microanalysis of biological specimens. *Scanning Electron Microsc.* 1980(2):309–320. (Biological quantitation)

Roomans, G. M. 1981. Quantitative electron probe X-ray microanalysis of biological bulk specimens. *Scanning Electron Microsc.* 1981(2):345–356, 344. (Biological analytical resolution, whole-sample quantitation)

Roomans, G. M., and A. Boekestein. 1978. Distribution of ions in *Neurospora crassa* determined by quantitative electron microprobe analysis. *Protoplasma* 95:385–392. (Biological sample preparation)

Sigee, C., and L. P. Kearns. 1981. X-ray microanalysis of chromatin-bound period IV metals in *Glenodinium foliaceum*: A binucleate dinoflagellate. *Protoplasma* 105:213–223. (Chromosomes)

Sild, E. H., and S. Pausak. 1979. Forensic applications of SEM/EDX. *Scanning Electron Microsc.* 1979(2):185–192. (Forensic applications)

Statham, P. J. 1984. Accuracy, reproducibility and scope for X-ray microanalysis with Si(Li) detectors. *J. Physique* 45(C2):175–180. (Instrumentation)

Sumner, A. T. 1978. Quantitation in biological X-ray microanalysis, with particular reference to histochemistry. *J. Microsc.* 114:19–30. (Biological quantitation)

Sumner, A. T. 1983. X-ray microanalysis: A histochemical tool for elemental analysis. *Histochem. J.* 15:501–541. (EDS in histochemistry)

Williams, D. B. 1987. *Practical Analytical Electron Microscopy in Materials Science.* Philips Electronic Instruments, Inc., Electron Optics Publishing Group, Mahwah, NJ. (Cliff-Lorimer quantitation)

Zeichner, A., H. A. Foner, M. Dvorachek, P. Bergman, and N. Levin. 1989. Concentration techniques for the detection of gunshot residues by scanning electron microscopy/energy dispersive X-ray analysis (SEM/EDX). *J. Forensic Sci.* 34:312–320. (Gunshot residue detection)

Ziebold, T. O. 1967. Precision and sensitivity in electron microprobe analysis. *Anal. Chem.* 39:858. (Statistical analysis)

Zierold, K. 1983. X-ray microanalysis of frozen-hydrated specimens. *Scanning Electron Microsc.* 1983(2):809–826. (Frozen-hydrated specimens)

9

Electron Micrographic Techniques

One of the most important aspects of any type of microscopy is the ability to record an image of what is viewed. If results are to be discussed and disseminated, this step is essential in the scientific use of the instrument. Early on, in light microscopy, the microscopist relied on unbiased digital transduction (drawings) of what was seen, and although this method does allow the microscopist to filter noise from the image, subjectivity can creep into the final product.

In the second half of the nineteenth century, photographic equipment that could be mounted on a microscope became available, greatly reducing the subjectivity of the recording process. By the time the prototype electron microscopes were built, the need for direct recording capability was recognized and incorporated as an integral part of the initial designs.

Electron microscopes have changed greatly in the last half of the twentieth century and now are capable of recording much more of the electron–specimen interaction than just morphological information. The ability to produce a lasting image of a small and sometimes ephemeral object, however, is still a major objective of electron microscopy.

EM imaging systems and their peripheral equipment can record electron images or transduced light-images of electron interactions. There are two broad approaches: traditional photographic methods and electronic recording means.

SILVER GRAPHIC PROCESS

Photography, in its most basic and traditional sense, is the capture of a luminous image by creating a metallic silver image. Regardless of the imaging medium, the fundamental mechanism of the silver-halide-based graphic system is essentially the same. In the cases of light micrography, the micrographic process in the SEM, and micrograph printing, the information-carrying energy source is light (photons), whereas in TEM, it is electrons. The graphic process is generally accomplished in two steps, with the initial image being recorded on an emulsion containing silver halide crystals supported by a transparent carrier called the negative. The transparent carrier may be either a sheet of glass or, more commonly, pieces of flexible polyester film. For conventional pictures (prints), this negative image is photographically projected onto a second sensitive emulsion which is supported by a highly reflective paper (Figure 9.1). This process reverses the image and yields a positive print.

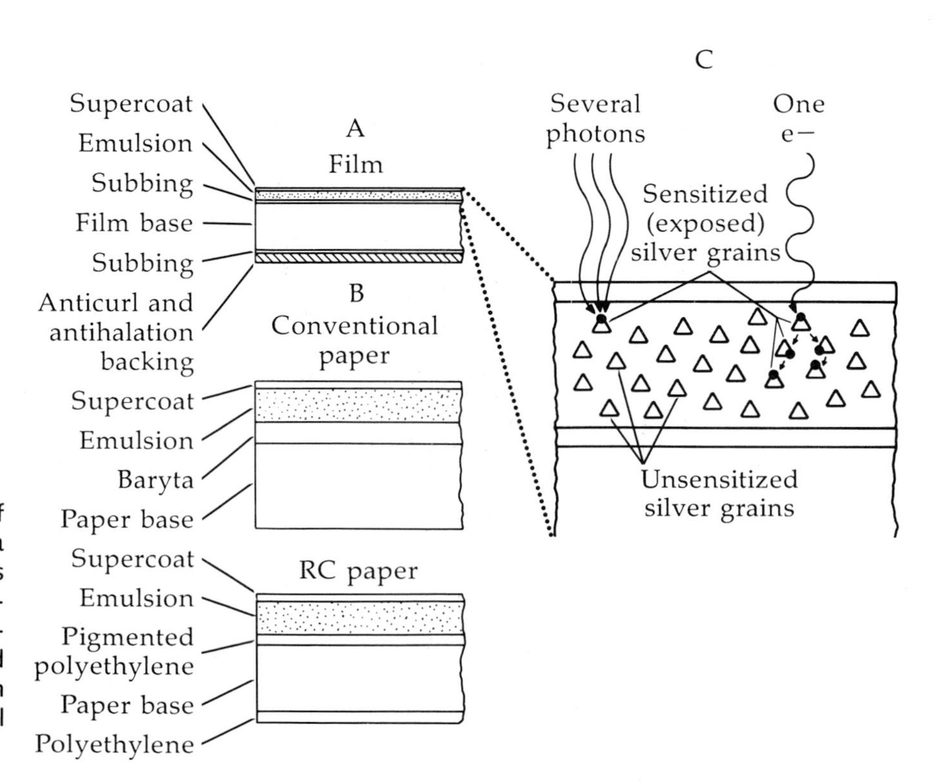

Figure 9.1 Composition and action of sensitive materials. (A) Schematic of a cross section of a typical film. (B) Cross sections of coated and noncoated photographic paper. (C) Schematic of the interaction of light or beam electrons and photographic emulsion. A single beam electron has far more sensitizing potential than a photon does.

Latent Image Formation

The image generated by these means is composed of thin deposits of metallic silver. These deposits are formed by the reduction of silver halide crystals in the emulsion that have been sensitized by one of the energy sources mentioned. For most purposes, the silver halide crystals have dimensions in the range of 30 nm to 2,500 nm and are applied along with other ingredients to the support medium in a gelatin film or emulsion.

The sensitization of the silver halide crystals (grains), which is required for their preferential reduction by the developer, results from the neutralization of silver ions within the crystals either by direct collision with imaging electrons or by interaction with photo electrons generated by a light source. If visible light is the sensitizing agent, where the photon energy levels are less than 10 eV, several quanta are needed to add enough energy to form a grain that can be developed. On the other hand, electrons from the imaging beam in the microscope have energies of at least 20,000 eV, enabling them to sensitize the silver grains easily but bringing about some other problems (discussed later).

The migration and aggregation of three to four neutralized silver atoms within the crystal should form a developable grain, but in practice, about ten seems to be the threshold. The distribution and density of these developable grains across the graphic medium form a latent (invisible) image (see Figure 9.1). Since the image is invisible and can slowly gain overall density during storage, it is better to stabilize it by chemical development as soon as possible.

Chemical Processes Reveal and Preserve the Image

Once a negative has been exposed and contains a latent image, the chemical processes to yield a visible and stable image are the next steps (Figure 9.2). The objective of the first step (developing) is to reduce the sensitized silver halide to metallic silver without reducing

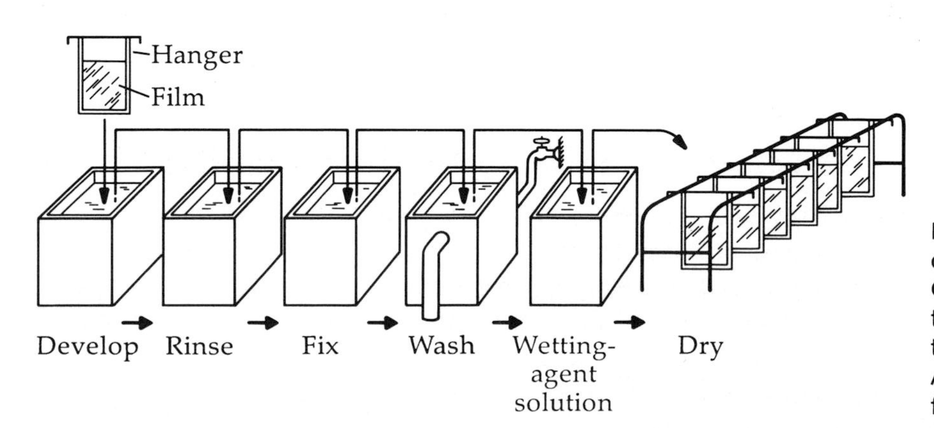

Figure 9.2 Processing steps in the development of a photographic negative. Controlling temperature, time, and agitation control during developing is crucial to obtaining uniform negative quality. Automated processing systems greatly facilitate the procedure.

the unsensitized surrounding material. This process is accomplished, under a safelight, by submerging and agitating the negative in a bath containing reducing compounds, such as a combination of hydroquinone and p-methylaminophenol sulfate (Metol) in an alkaline solution. With the proper conditions of temperature and development time, this step leaves a visible silver image embedded in the film emulsion. If the negative were left in contact with the developing solution for a protracted period and/or the temperature were too high, the developer would proceed to reduce even the unsensitized silver grains, eventually leading to a completely blackened emulsion with no image.

After the negative has been developed to the desired density (image intensity), the next part of the processing sequence is to remove the undeveloped silver halide crystals so that the silver image can be viewed under room lights. This step is called fixing. To get the most use from the fixing bath and to ensure uniformity in the negative, the residual developer is either washed off the negative with running water or it is neutralized in an acidic (usually acetic acid) stop bath. The choice of method depends on the characteristics of the negative emulsion. For the common TEM emulsions, a water wash is recommended.

Fixing baths generally contain sodium or ammonium thiosulfate (hypo) as the silver solvent. This material forms complexes with the remaining silver halide in the negative emulsion, rendering it soluble, and enabling it to be washed out, leaving the permanent silver image. Depending on the exact sequence of further processing, the fixer may also contain agents designed to toughen the emulsion gel to protect it from scratches and other mechanical damage.

Following fixing, the negative is washed in running water for a period long enough to ensure that all of the fixing agent has been removed. If this were not done, the residual sulfur from the fixer would eventually form a yellowish brown silver sulfide that would tarnish the silver image, causing it to fade. In addition to a running water wash, many microscopists use a bath of a hypo-removing agent to ensure the complete removal of the fixer from the negative. After this wash, the negative can be dipped briefly into a wetting agent (to prevent water spots), then air-dried in a dust-free location.

Exposure and Developing Determine Image Quality

The image that results from this graphic process is dependent on a number of characteristics of the exposure medium, the film emulsion, and the processing conditions. In general, the goal is to preserve spatial (morphological) and tonal (electron density) information. The first category is largely dependent on specimen preparation and microscopic techniques; the latter type also depends on proper manipulation of the graphic medium.

The degree to which a negative is exposed is related to the total number of photons or electrons that hit it and thus is related to the product of the exposure time and the brightness or beam current of the source. As the exposure increases, the number of sensitized silver grains increases (hence an increase in the number of silver atoms reduced upon development). The amount of reduced silver in the emulsion is directly related to the optical density of the negative. This is more easily interpreted over the potential range of densities when the exposure scale is presented logarithmically. The resultant curve is called the characteristic curve for the film (Figure 9.3).

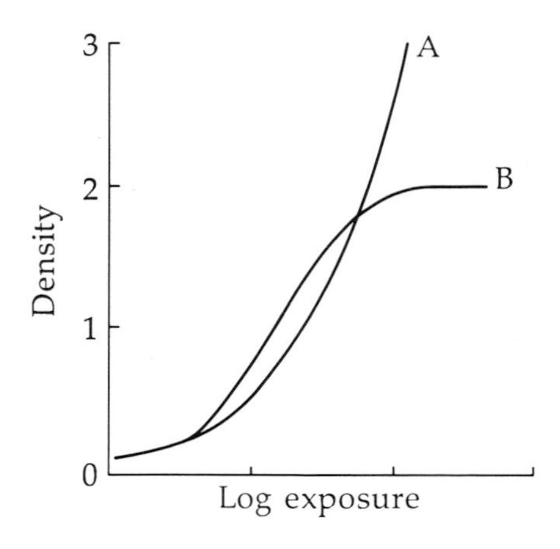

Figure 9.3 Characteristic exposure curves. (A) Emulsion exposed to electrons. The contrast (change in density/log relative exposure) increases continuously with increasing electron exposure but is constant over much of the emulsion range for light. (B) Photographic film exposed to light.

It is useful to consider the relatively linear portion of this curve. The effect of changing the slope of the line that it approximates (i.e., the change in density over the change in log exposure) is to change the gamma or contrast of the film or paper. As the slope increases, a smaller difference in exposure is needed to yield a given change in density. Thus, for an image with a wide range of brightness values, a lower slope (lower gamma) would allow more information to be captured, since the density change would be spread over a wider range of exposure. Conversely, if the image to be recorded had a limited range, a high-gamma film would allow a greater range of densities on the negative.

Film contrast is also affected by the activity of the developer, so a given film can produce a range of contrasts, depending on how it is developed. Longer development times, higher temperatures, and more agitation during developing all increase the contrast of the negative. When developing is carried to the extreme, however, the problems of graininess and background fog become detrimental to the image.

PHOTOGRAPHIC PRINTING

Just as the initial recording process of the negative uses the silver imaging process to capture the specimen image, photographic printing uses the same approach to reveal the information contained in the negative in a sensible manner for communication. The micrograph print is the fundamental means of showing EM results.

Contact Printing and Enlarging Are Basic Approaches to Printing

The simplest form of printing is the contact print. Making contact prints usually is a worthwhile exercise with any series of negatives, since contact prints provide a record of the negatives that can be viewed easily and directly. To make the contact print, the negative is simply

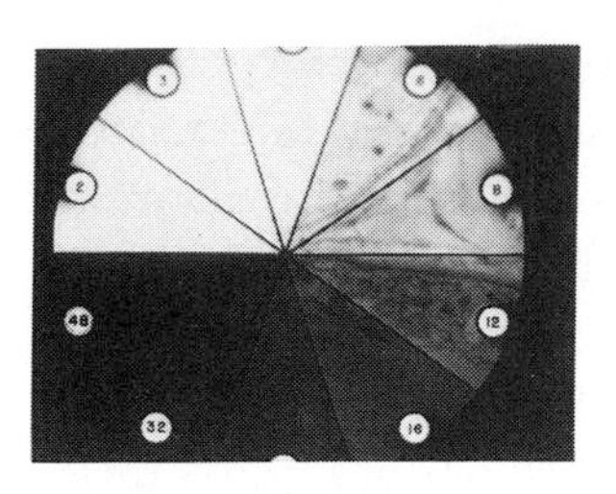

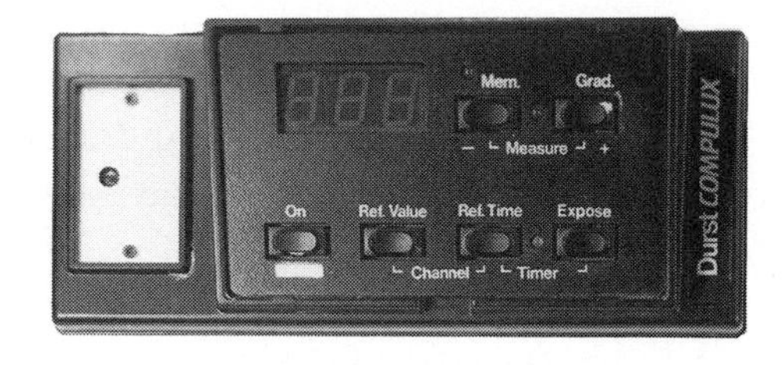

A B

Figure 9.4 Methods of judging correct print exposure. (A) Test print, produced by overlaying a step filter calibrated directly in seconds. (B) Hand-held densitometer. This unit acts as an interval timer as well and can recall several paper speeds for printing at different contrasts.

placed (emulsion-side down) in tight contact with a piece of photographic paper, and the paper exposed, through the negative, by a uniform light source. The light can be from almost any source, but usually a photographic enlarger is used if there is no dedicated contact printer available.

The correct exposure is judged either by making a test print with a series of short exposures done with a corresponding series of partial masking, or by evaluating the negative density with an electronic densitometer (Figure 9.4). Once exposed, the test or initial print is subjected to a wet-chemistry regimen similar to that used on the negative (the theory is identical).

The human eye has a resolving power of about 200 μm, and the silver grain size in the negative emulsion is at least 40 times smaller than this. Thus, considerable optical enlargement should be possible without producing simply empty magnification. Although the greatest useful magnification may not be the theoretical 40 times, generally 8 to 10 times is practical. This magnification is achieved by projecting the negative image onto the printing paper with a photographic enlarger.

Figure 9.5 Standard large-format enlarger. This straightforward, rugged design is ideal for EM printing. Numerous control options are available from the manufacturer to improve printing efficiency. The enlarger projects the negative image onto photographic paper to produce a magnified, positive print. (A) Lamp housing and condenser lens system. (B) Negative carrier. (C) Enlarging lens.

Photographic Enlargers Restore Hidden Resolution

Photographic enlargers have three functional subunits: an illumination source, an optical projection system, and a means of supporting and adjusting the imaging components (Figure 9.5). The light source generally falls into one of three categories depending on the approach used to illuminate the negative (Figure 9.6).

Point-source enlargers use a small, high-intensity lamp above a series of condenser lenses to provide the most uniformly collimated illumination of the negative. These light sources give the image the highest apparent contrast, which is often necessary with the typically low-contrast TEM negatives produced from photographing biological samples. They also require that the user keep the negative surface scrupulously clean because the highly collimated light source accentuates any scratches or dust particles. These light sources also generate considerable heat, which must be drawn off, usually by a remotely mounted fan. Care must be taken to see that no vibrations are set up in the enlarger by this apparatus.

Condenser light sources illuminated by a larger, diffuse lamp are a type commonly used in darkrooms. These enlargers employ a series of condenser lenses below a large frosted bulb to provide uniform light to illuminate the negative. They also provide a highly collimated illuminating beam, although not quite as vertically oriented as the point-source system because of the wider bulb. Again, these light sources give a high-contrast image and are also sensitive to scratches and dirt on the negative or in the optical system. They are more forgiving of minor misalignments than point-source condenser illumination systems are.

A final category of light source for enlargers is the diffusion type. These illuminate the negative with a luminous field above it. Diffusion-type light sources provide the most uniform illumination, but the light rays are of a diffusive rather then collimated nature. These light sources generally do not provide as high a level of apparent contrast as the condenser or point-source types, almost a 20% difference in contrast in some applications. Diffusion sources are more often used for printing conventional images, in which the negative has considerable inherent contrast. The illumination for these sources can be provided either by conventional tungsten-filament bulbs or by a fluorescent cold-light source. The latter gives very even illumination with little heat input to the negative.

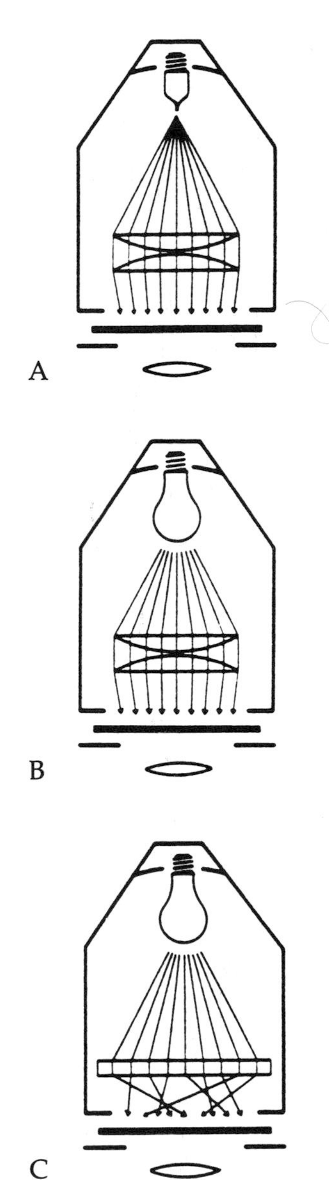

Figure 9.6 Illumination methods for photographic enlargers. (A) Point source with condenser-lens system. (B) Diffuse source with condenser lens. (C) Diffuse source. These sources are pictured in order of decreasing contrast. Diffusion sources, the most forgiving of flaws in the negative, generally provide too little contrast for electron micrographic work.

The main optical component of the enlarger is its lens. The lens ultimately is responsible for the amount of detail extracted from the negative. Buying the best quality that can be afforded is generally appropriate, since even the best lenses, with proper care, will last indefinitely. A lens with a focal length near the normal photographic-lens focal length for the negative format being processed generally gives the best performance for EM work. Thus, for 35-mm format films a 50-mm lens works well, and for the common 3.25-inch by 4-inch TEM plate films and Polaroid films a 135-mm or 150-mm enlarging lens works well. Each lens should have an adjustable iris diaphragm; a shutter is useful for contact printing.

The mechanical frame of the enlarger can be one of a number of designs. The design has little effect on the image quality, except in terms of how susceptible the support is to vibration. In this regard, wall-mounted systems are probably the most stable, followed by rigid bench types. Since most EM enlarging is straightforward (i.e., no optical tilts or image shifts are involved) and is in the range of 1.5× to 5×, a simple enlarger with a sturdy frame is probably the best choice.

Once the negative has been mounted in the enlarger and the desired degree of enlargement selected, the image is focused and formatted on the easel or printing-paper carrier. Here a reflectance-grain focuser, that allows the user to focus on the emulsion grain (i.e., the exact image plane), is very useful. A test print is then made of the negative to determine the correct exposure time for the printing paper to be used, as was done for contact printing.

Despite the best efforts of the microscopist to provide a properly exposed negative with optimum contrast, sufficient variation occurs to require control of image contrast at the printing level. Print contrast control can be approached in two practical ways for scientific printing. First, the microscopist can use serially graded papers that have similar finish and processing characteristics, but have emulsions with different slopes in their characteristic curves. Alternatively, a single multigrade paper may be used. Multigrade papers have an emulsion that is sensitive to both the quantity *and* the quality of light used in the exposure. Filtering the enlarger light (either before or after it interacts with the negative) enables a variety of contrasts to be printed from the same paper.

As mentioned earlier, the chemical processing of the exposed print is similar to that described for the negative. Quite often the only changes are in developer composition and developing times. A series of processing trays is generally used in the EM darkroom, with the print being transferred manually from solution to solution with soft-tipped tongs. Considerable time can be saved in print processing if resin-coated papers are used. This concept limits the chemical exposure of the print to a few micrometers outside of a plastic coating, which keeps the paper fibers from becoming saturated with processing chemicals (see Figure 9.1) and decreases the processing time from hours to minutes. The glossy coatings also eliminate the need for ferrotyping the print (drying it against a heated metal mirror) to achieve a high-gloss finish.

Although the resin-coated papers do not have quite the image quality or archival properties of the finest fiber papers, they are quite adequate for scientific purposes. In addition, their short processing time makes them compatible with automated print processors, which can handle the whole wet process automatically, further reducing darkroom time.

TRANSMISSION ELECTRON MICROGRAPHY

As mentioned in the beginning of this chapter, even the first TEM had provisions for recording the image graphically. Almost all modern TEMs, regardless of what other information-detection equipment they may have, are equipped with at least one built-in camera system, which is the primary means of transducing specimen structural information.

TEM Micrographic Imaging Uses Three Systems

TEM micrographic imaging systems are generally composed of three subsystems: a viewing/focusing screen, an exposure metering system, and the camera system (Figure 9.7).

The focusing screen in the TEM is either part of the main screen or is a smaller screen inserted into the beam path. These screens are coated with a phosphorescent coating that radiates light after being struck by an imaging electron. Since the focus of the TEM varies with the magnification setting of the intermediate lens, the image must be focused (by compensating with the objective lens strength) at the final magnification. This process can be aided by the use of a binocular microscope focused on the phosphorescent screen. Another useful focusing feature is a deflecting coil that rocks the beam off of its axis. If the image is in focus, no apparent lateral displacement will occur as the angle of the beam changes. If it is not, a lateral shift or wobble will be apparent in the viewed image. If the image appears sharp through the binoculars, which generally magnify about 10 times, it will probably remain sharp at any reasonable photographic enlargement.

Another part of the imaging system is the exposure meter. It functions either by measuring the light emitted by the viewing screen, as a conventional photographic light meter does, or by measuring the beam current flowing to ground with a sensitive ammeter. Since this current is directly related to the beam intensity, an exposure time can be calculated easily. The latter approach is what is found on modern machines.

The exposure value given by a screen metering system represents a reading for the average of the screen. If the sample is relatively uniform in its coverage of the screen and is of a reasonable range of

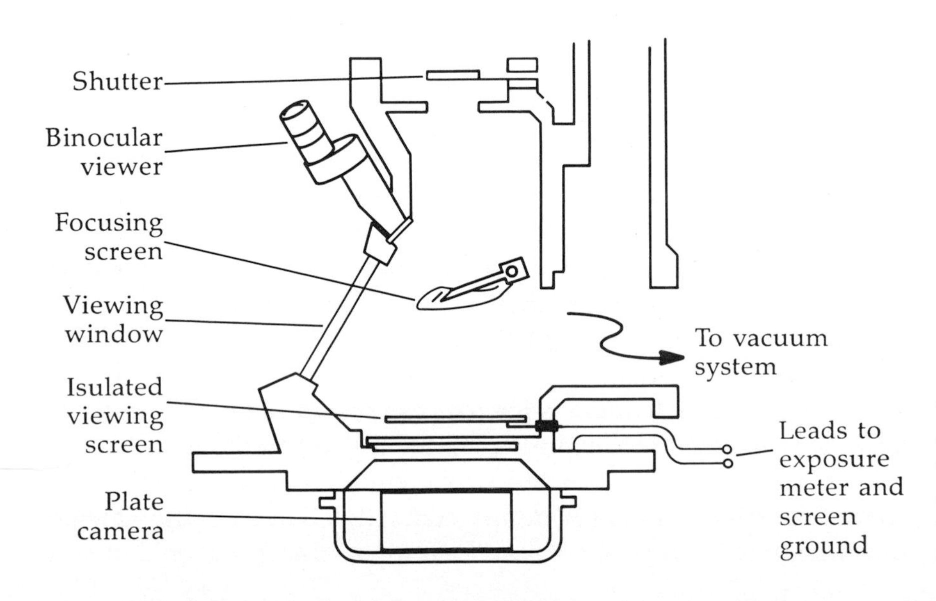

Figure 9.7 Schematic section through a TEM camera chamber. The great depth of focus of the TEM projector lens allows the focused image to be viewed at several points. The image will still be in focus on the plates, which are exposed at a level below the viewing screen.

values, this reading will produce an acceptable negative. If the area to be photographed has either grid bars or holes filling a significant part of the field of view, however, the exposure must be adjusted so that the negative density of the area of interest can be printed. Diffraction patterns are similarly difficult to judge by average screen current.

The final component of the TEM imaging system is the camera itself. The camera is an integral part of the microscope and consists of a mechanical shutter (open during viewing) and a film transport system. Most micrographs taken today are done on single-sheet film negatives that are transported to the imaging area, usually directly under the viewing screen, by electromechanical or pneumatic means.

In general, the exposure process is relatively automatic on modern machines. At the push of a button, the film is advanced to the imaging area, the beam is stopped by the shutter, and a part of the viewing screen opens to allow exposure of the plate. The shutter then opens for a predetermined time, exposing the plate while an optical system records the exposure, magnification, and serial number on the edge of the film. (Older machines may lack some or all of these last features.)

Finally, the plate is transferred to a receiver box (all still under vacuum) for storage until processing. Most microscopes have provisions for taking 15 to 50 such pictures before needing to be reloaded.

Electron Sensitometry Differs from Light Sensitometry

The microscopist can control a number of variables that affect the TEM image. The objective in manipulating variables should be to take quality micrographs that record information efficiently. Since the beam electrons can sensitize a number of emulsion grains, the appearance of graininess in an evenly illuminated area can be due to randomness in the impact of the imaging electrons with the film surface. This graininess is more pronounced than in a light image given equivalent exposure.

The first approach to limiting graininess is to increase the signal-to-noise ratio of the exposing beam by increasing the total exposure. This can be an increase in time or in intensity, since the film response is a quantum event. For most biological systems, the operator needs to balance the exposure time and intensity for practical reasons. Biological samples are prone to movement, degradation, and drift induced by the electron beam at high intensities, which would indicate that a long exposure time (lower intensity); therefore, less heating would be desirable.

Another consideration is the focus setting of the final condenser lens. As mentioned in Chapter 4, for the highest resolution, the lens should be defocused to provide a coherent illuminating beam to the specimen. This may not provide an electron flux-density at the viewing screen high enough to allow easy focusing of the specimen and will require long exposure times. In general, if the specimen is properly prepared, an exposure time in the range of 1 to 4 seconds should provide enough electrons (approximately 1 $e^-/\mu m^2$) to expose the film in a conventional TEM with a tungsten filament properly. Images of sensitive samples can be focused on an area other than that of interest with a more highly focused condenser and then recorded by shifting either the sample or the beam.

If the calculated exposure time for a sample image appears to be excessive, an alternative is to lower the microscope magnification and regain the image size by photographic enlargement. Lower magnification, at a given condenser setting, allows more of the beam electrons to strike the screen or film and also allows the production of a negative of suitable contrast with a lower electron flux to the specimen. Although negative contrast increases with this method, the graininess of the negative also increases somewhat.

Even when not required for increased contrast, lower magnification is still worth thinking about when composing a picture. With most biological material, the negative need not be exposed at the final magnification at which it will be viewed. Resolution is determined by the objective lens, and a lower intermediate-lens setting provides a wider field of view, more total information, and flexibility in orienting and formatting the final print (Figure 9.8).

If there is no way to get sufficiently high electron doses to the film to yield a printable image, the microscopist can, as a final approach, increase the development time of the negative. Increasing the development time causes more aggregation of silver grains in the emulsion, however, leading to a still grainier image, which already is likely to be grainy because of the random electron-flux effects described earlier. Different films respond differently to this technique. The normal development on some may be so near to complete development of sensitized silver grains that increased development time only increases the background or fog density.

The procedure for processing TEM negatives is the same as for the general case for negatives described earlier. Since all light emulsions have the same electron speed (owing to the vastly greater energy of the sensitizing sources), electron imaging films can be made of slow, fine-grained emulsions that are sensitive mainly to blue light, which allows the whole procedure to be done under yellow safelight conditions. Low-pressure sodium-vapor safelights that emit nearly monochromatic yellow light (589 nm and 589.6 nm) and allow a light level in the darkroom greater than that in most expensive restaurants are available.

For much biological work, the image from the microscope is low in contrast compared to the contrast range of ordinary photography. To capture the full range of differences a high-contrast (high hydroquinone-to-Metol ratio) developer is usually used. Ordinary D-19 developer gives excellent results with all currently manufactured TEM

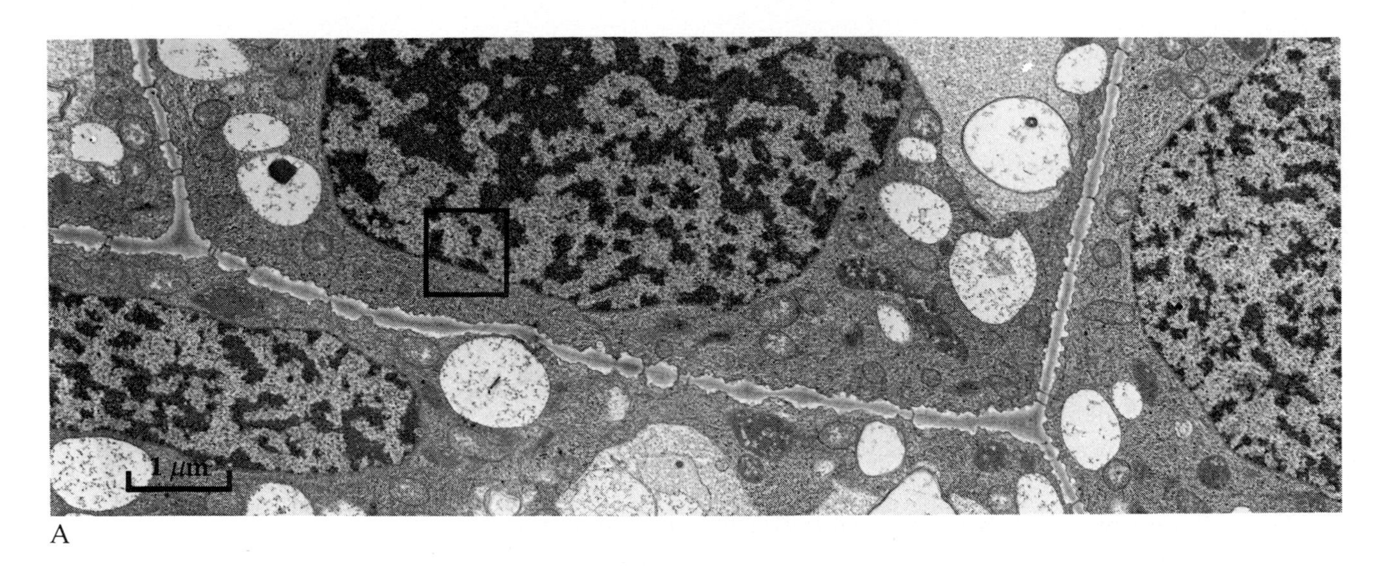

A

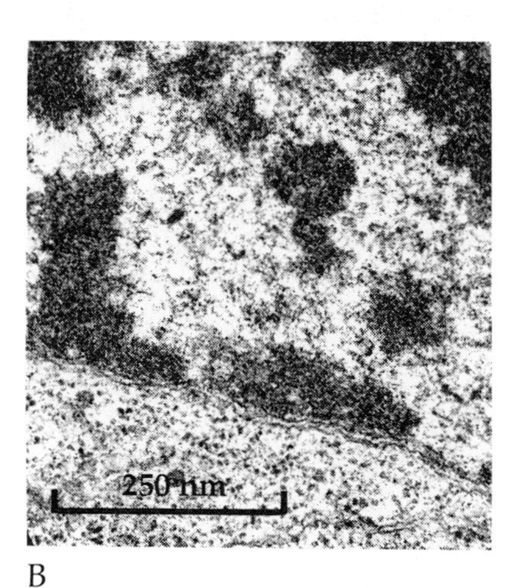

B

C

Figure 9.8 Micrograph magnification selection. (A) Low-magnification micrograph of an onion root cell. (B) An area along the nuclear envelope enlarged optically from this negative. (C) An electron-optical contact print of the same region at the same magnification. It is generally more efficient to use a microscope magnification that is lower than the desired magnification and bring the image to final size by enlarging it photographically.

negatives and can be purchased, prepared, or compounded from common laboratory reagents. It can be used at various concentrations, depending on the film type and exposure conditions. To achieve a high degree of uniformity in TEM negatives, uniformity in the development step is critical and generally should be held constant, with other conditions varied around it. Uniform temperature can be provided by a controlled-temperature jacket around the developer tank and uniform agitation by a timed nitrogen-burst system. Careful attention should be paid to development time and to the age and capacity of the developer.

Printing TEM negatives is operationally no different from printing any other medium-format negative. The films currently produced have plastic bases that are rigid enough to preclude the need of glass negative carriers (thereby eliminating four more surfaces in the image plane that can attract dust). If rational consideration is given to the most efficient microscope magnification, considerable latitude in final print enlargement and orientation should be possible.

SCANNING ELECTRON MICROGRAPHY

SEM Micrography Is Done with Light

Micrography on the SEM or STEM, in contrast to the TEM, is a strictly photographic process. As described in Chapter 5, the SEM image is viewed on a CRT. Viewing the image in this manner allows the user to make wider changes in image brightness, contrast, and orientation than are possible with the TEM.

The image-recording CRT in the SEM has rasters of much higher density and shorter duration than the rasters of conventional television tubes. The photographic raster of sweep is composed of about 2,500 lines, each slightly overlapped, yielding a meaningful final screen magnification size of about 25 cm (or about an 8-inch by 10-inch print). Since the final image-recording system in conventional scanning machines is photographic, the user has a range of imaging materials available. Although provisions can be made for the use of conventional sheet-film or roll-film cameras, the instant results from Polaroid materials are usually desirable.

Polaroid Films Produce Instant Images

Polaroid films are available in several formats, speeds, and contrast grades and come with or without a printable negative. In general, the positive/negative (P/N) systems are the most desirable for scanning work, since along with the print they provide a printable negative, which gives the user some latitude in the contrast, size, and format of the final print. The P/N negatives also have a resolution limit about ten times that of the contact print formed in the Polaroid process. The P/N contact prints must be coated with a protective coating (provided in the package) to keep the image from fading. The slower speed of the P/N films (compared to that of films that produce only positive images) has no effect on their use in SEM; exposure time is determined by the intensity of the imaging scan.

Polaroid film processing is started when the exposed film packet is pressed through a roller system in the Polaroid film pack, causing a packet of developer material to burst and be smeared between the negative and the positive receptor paper. Sensitized silver halide grains in the negative layer are then reduced to metallic silver, and the unexposed silver halide migrates into the receptor paper, where it complexes at certain catalytic nuclei and is reduced to form a positive silver image (Figure 9.9).

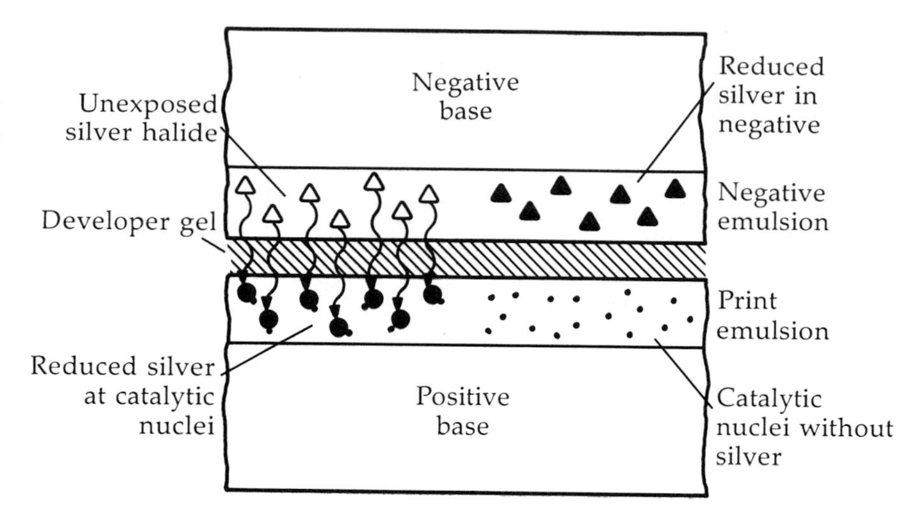

Figure 9.9 Basic method of Polaroid film action. Silver leached out of unsensitized areas in the negative by the processing gel migrate into and react with catalytic nuclei in the emulsion of the positive receptor layer to yield a print.

At room temperature, the development process is complete in about 20 seconds. There is some latitude for altering the contrast by increasing or decreasing this time, but achieving a balanced image in the microscope imaging system is usually more productive.

At the end of the prescribed development time, the film packet is peeled apart to yield a print and a negative. The negative must be cleared in an agitated bath of sodium sulfite solution to remove certain dyes, as well as the developer layer, and to soften the opaque backing layer. This bath is followed by washing, a wetting-agent bath, and drying (Figure 9.10). Once processed, Polaroid negatives can be stored and used to prepare enlargements in the same way as the TEM negatives described earlier.

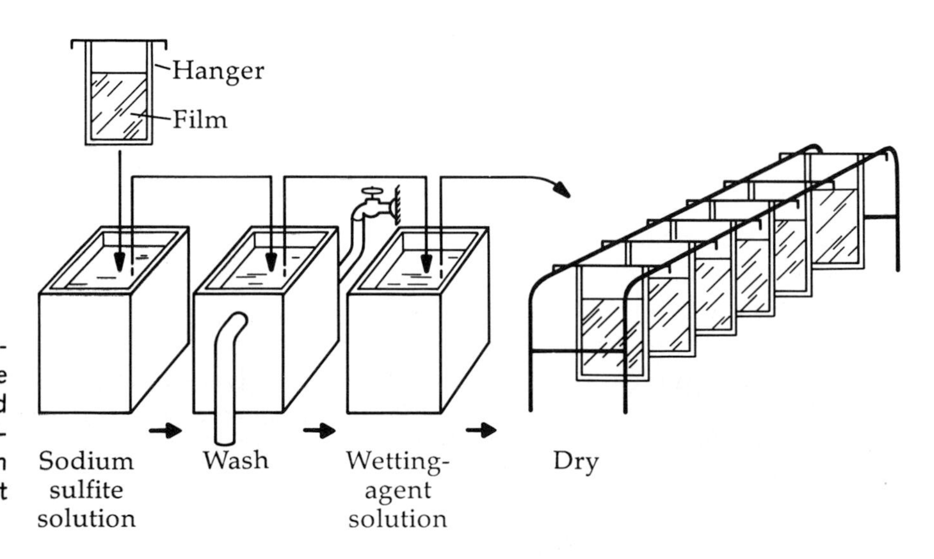

Figure 9.10 Processing sequence for Polaroid negatives. After developing, the residual developer needs to be removed from the Polaroid negative and the emulsion hardened. The negative is then washed and treated with a wetting agent prior to drying.

THE ELECTRONIC DARKROOM

Since the inception of electron microscopy, the standard method of recording EM images has been the silver-based photographic method. This aspect of electron microscopy continues to be one of the most time-consuming and costly. Rapidly emerging technologies, however, may limit the role of traditional photographic techniques in the EM laboratory. These advances are being made at all levels of postgeneration image manipulation and may soon limit, perhaps even eliminate, the need for the photographic darkroom.

Newer Methods of Image Capture and Processing

Rather than recording the TEM image on a sensitive film and optically transducing the image to photographic paper, direct handling of the image electronically is now possible. The general approach is to move into the beam a scintillator (single crystal), in place of a viewing screen, and allow the resultant light image to be piped to a charge-coupled device (CCD) either directly in the microscope or in an externally mounted camera (Figure 9.11). This digital image can be transferred to an image-processing board in a microcomputer and thus made available for display, further processing, storage or printing.

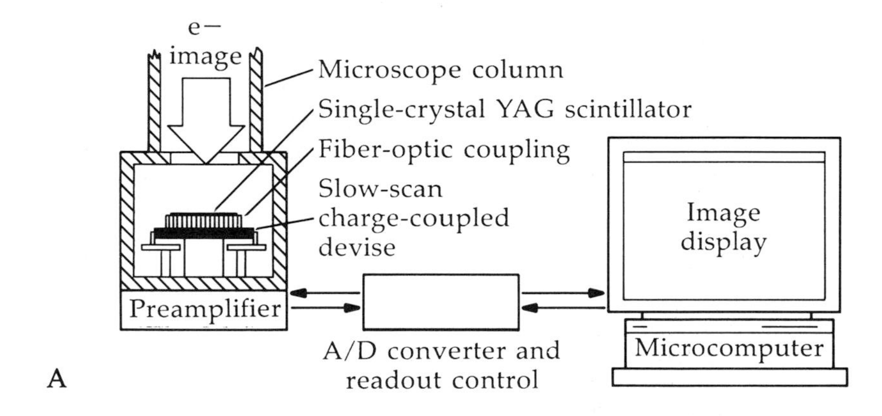

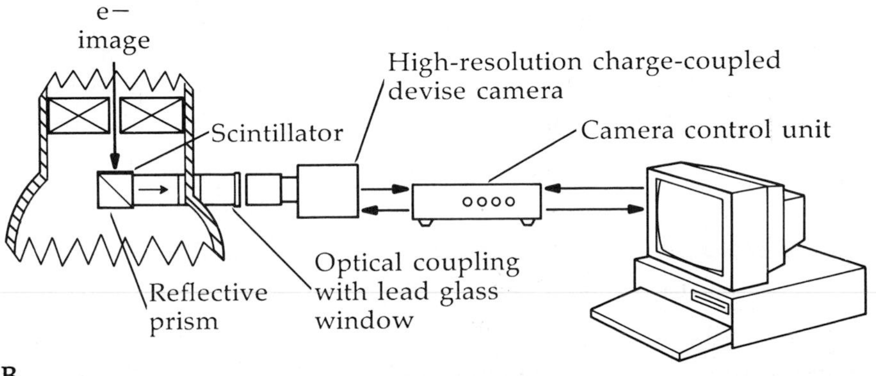

Figure 9.11 Digital cameras for TEM image capture. (A) An internally mounted slow-scan CCD device. (B) An externally mounted high-resolution CCD camera. The internal unit is more sensitive; the external unit is considerably less expensive. Part (A) redrawn with permission from Gatan, Inc.

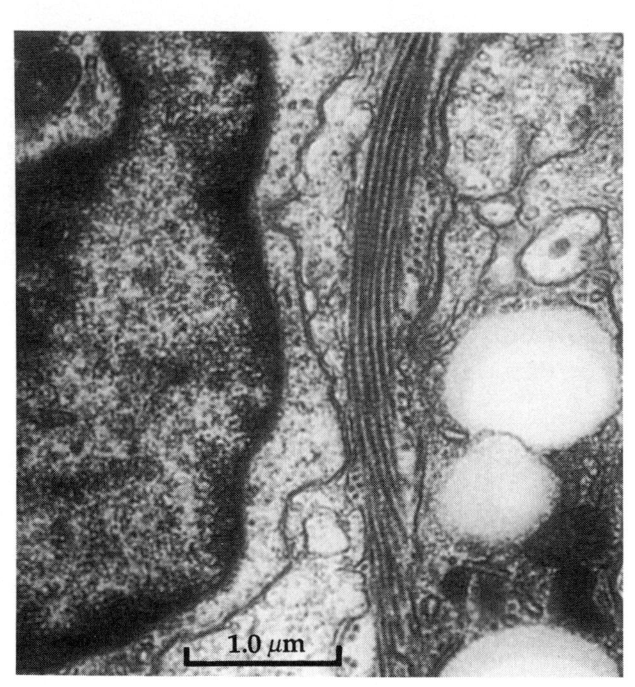

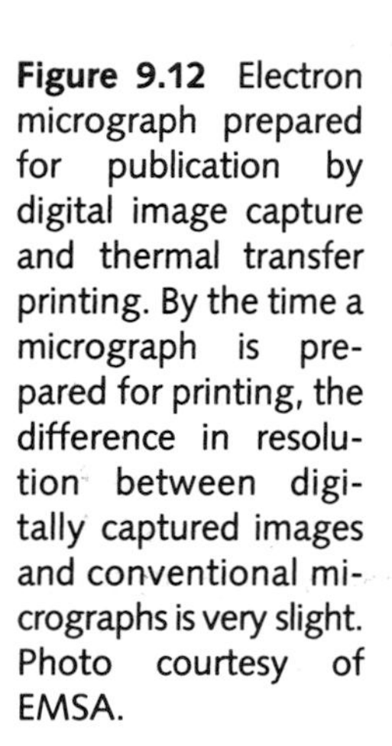

Figure 9.12 Electron micrograph prepared for publication by digital image capture and thermal transfer printing. By the time a micrograph is prepared for printing, the difference in resolution between digitally captured images and conventional micrographs is very slight. Photo courtesy of EMSA.

The CCD devices for imaging available at present have densities of about 1,000 by 1,000 picture elements (pixels). The TEM films are equivalent to about 10,000 by 10,000 pixels, making choice of image size and composition for the electronic medium a little more complicated. The CCD, however, has a higher dynamic range and no granularity noise, as compared to film, yielding a an image of higher quality on a smaller field. The hardware is improving, and densities at least four times greater than the 1,000-by-1,000 arrays now available are being developed.

Rather than the traditional darkroom methods for printing, the digital image can be stored on some mass storage device (e.g., a disk) or transmitted to a printer (Figure 9.12). Depending on the information needed, the printer can be a dot matrix printer, laser printer, or, for the highest resolution (publication quality), a thermal transfer printer. These devices are capable of producing a high-quality, relatively permanent copy (life of about five years at room temperature) in a matter of seconds. The whole time between seeing an image to record and having a hard copy of it can be reduced from hours to a few minutes.

Electronic Image Capture Speeds Analysis

In many modern TEM investigations, the user requires numerical data from the microscope image. Examples include ultrastructural dimensions, particle counts, differential stain reactions, reconstructions, and image filtering. Traditionally, these data have been collected from a series of calibrated micrographs, which then must be measured, transformed, or digitized by hand. The direct recording of electron images for further processing in these kinds of studies is the real potential of the electronic darkroom. Computer programs exist that allow the quantitation and discrimination of the size of features in digital images and the rapid presentation of numerical data with statistical comparisons included. Development in this exciting area promises to add measurably to the utility of EM in both biological and materials sciences.

MICROGRAPH PRESENTATION AND PUBLICATION

Regardless of the EM medium, once the final print has been produced, measurements can be made, differences can be compared, and hypotheses can be tested. Since the scientific cycle is not complete until the information is disseminated, the microscopist still has the task of presenting the EM information to the scientific community. Ideally these steps are left to the care of a competent photographic technician, but in practice the responsibility often falls on the shoulders of the researcher.

A widely used format for presenting EM information to large groups is 2-inch-square projection slides. A complete lecture can be composed using a combination of graphics, text, and micrograph slides; when done properly, a more efficient method of information transfer is hard to imagine. Micrograph slides can be prepared in many ways, but the techniques fall into two groups: those that use the EM negative directly to form the slide image and those that incorporate a print of the original micrograph. The former method involves the photography or contact printing of the illuminated negative onto transparency material (usually a black-and-white 35-mm film). This process yields a positive image and should allow rendition of nearly the complete range of values in the negative. If the final transparency is made from a black-and-white print, the range of values is compressed because of the smaller density range possible with the reflected light image of a print.

A number of films can be used for the presentation transparency of an EM print. The best tonal ranges and contrast are usually achieved with continuous-tone, fine-grained, black-and-white film. Kodak T-max black-and-white film, for example, can be reverse-processed with a packaged kit to yield a positive image with excellent results. A number of direct-positive films are also available, but care must be taken to achieve a proper contrast with these products.

Perhaps the fastest method of obtaining slides for projection is to use color film. Although wasteful of the color potential of the film, slides can generally be obtained from any photo processing lab within hours of exposure and usually at a price far more economical than the time it would take an individual to accomplish the task in a conventional EM darkroom. Using color film requires that the photographer pay closer attention to the quality of light than is required for black-and-white film. If a film with the wrong color-temperature (i.e., daylight rather than tungsten-balanced) is used, the micrographs will all have an annoying overall color.

Whether preparing micrographs for prints or slides, internal markers are worthwhile in all but the simplest compositions. At the very least, a bar of some known length should be included. For example, with a micrograph print of known final magnification (10,000× microscope magnification and 2.5× optical enlargement, or 25,000× total magnification), it would be appropriate to place a scale bar 25 mm long in an unobtrusive location on the print. This distance would be equal to 1 μm in the print and would stay the same whether one looked at the print or at a greatly enlarged, projected image. If the only scale information given is the 25,000× of the initial print, the viewer has no absolute grasp of the size when the image is projected. Additional markers, such as arrows to points of interest, can also be helpful in directing the viewer's attention to important details.

When preparing plates for publication, the microscopist should understand the size requirements and limits as early in the process as possible. Generally space can be saved by combining more than one image in a plate. A useful technique in many applications is to use a lower-magnification background micrograph and include higher-magnification inserts of selected areas of interest. As is the case for projection slides, an internal size marker should be present. If the plate is for journal publication, the printer may alter the final size, making any statement of magnification in the text incorrect. Composing the plate at a greater scale than the final rendition is also useful in reducing the effect of assembly errors.

TEM shadowcast images, for any use, are generally easier to interpret if presented so that the lighter part of the image is toward the top of the page. In the case of a conventionally processed image, the lighter part is the shadow, but the mind is compelled to interpret images with light from above. Quite often shadowed images, especially those shadowed at acute angles, are easier to interpret if the image is reversed. This reversal can be even more important when viewing small pit or hill structures in an area that has few familiar structures. Micrographs that are printed photographically reversed have optically satisfying dark shadows (the specimen looks like it was illuminated by the shadowing metal) (Figure 9.13).

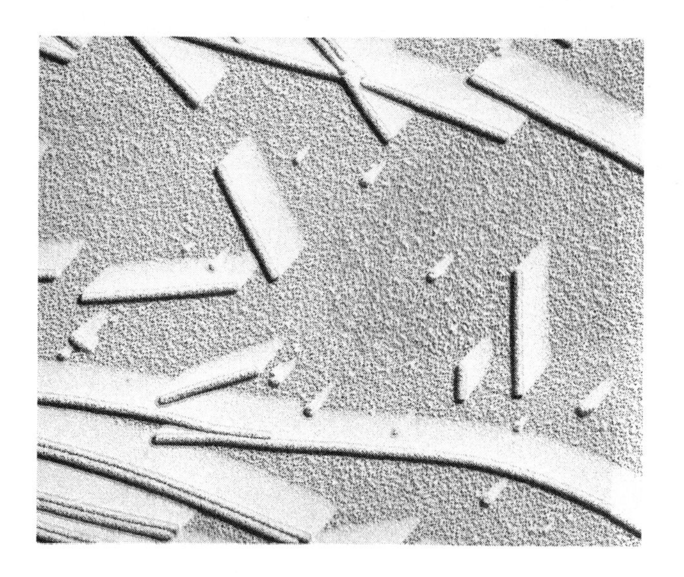

A

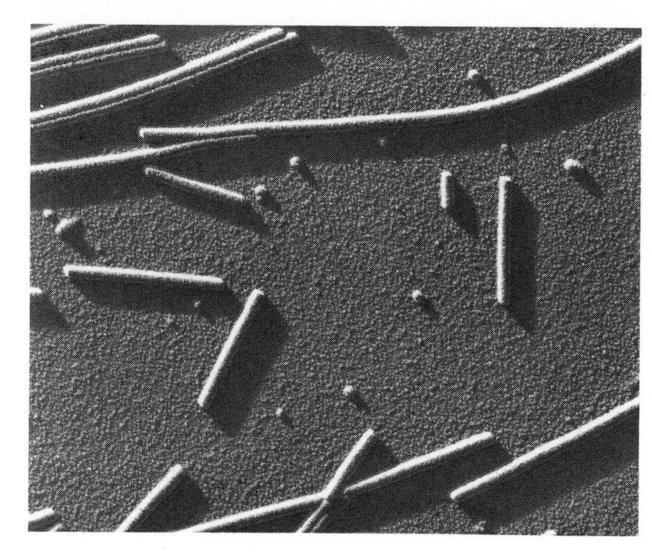

B

Figure 9.13 Image reversal for shadowcast micrographs. (A) Micrograph of tobacco mosaic virus printed directly. (B) The same image photographically reversed. By reversing the shadowcast image, the specimen appears to be illuminated by the shadowing metal. The shadows are rendered dark, which makes the image easier for most viewers to interpret than if the shadows are white.

FURTHER READING

Farnell, G. C., and R. B. Flint. 1975. Photographic aspects of electron microscopy. In *Principles and Techniques of Electron Microscopy*, vol. 5. M. A. Hayat, ed. Van Nostrand Reinhold, New York. (TEM imaging)

Jacobson, R. E. 1978. *The Manual of Photography*. Focal Press, London. (General photography, photochemical formulary)

Kodak Electron Image Plates. 1985. (Kodak pamphlet no. N-921.) Eastman Kodak Company, Rochester, NY. (EM film characteristics)

Using Electrons Effectively. 1984. (Kodak pamphlet no. M6-121.) Eastman Kodak Company, Rochester, NY. (TEM micrography)

INDEX